AF885428

SOTHIS PRESS

Sothis Press aims to completely reimagine the public domain book with an emphasis on universal access to captivatingly designed print editions. We believe that technology will enable every single book ever published to be perpetually accessible in beautiful print form. Sothis is beauty in ubiquity.

www.sothis.press
1 Waterhouse Square, London, ECIN 2ST

Shop and Foundry Management

Stuart Dean

SOTHIS PRESS

ISBN 978-1-0368-0976-8

Shop and Foundry Management
is in the public domain.

This edition is published by
Sothis Press © 2023. The company
is registered in England and Wales.
The Company Number is 14900389.

Shop and Foundry Management

BY

STUART DEAN
Superintendent Dean Bros. Steam Pump Works

NEW YORK
DAVID WILLIAMS COMPANY
231-241 WEST 39TH STREET
1913

PREFACE

FOR the most part this volume consists of articles written for The Iron Age, Mr. Dean's purpose being to present everyday shop and foundry methods that have resulted in lower cost and greater output. As superintendent of the Dean Brothers Steam Pump Works, Indianapolis, Ind., in the past thirteen years, the author has constantly aimed at four things: Reduction in cost of production; increase in plant capacity through greater efficiency; quick deliveries, and a perfect product.

Mr. Dean, who is nephew of the proprietors, was put in charge of the Indianapolis plant thirteen years ago at the age of 25. The product is pumps ranging from 24 to 50,000 pounds, some of them designed by the author, who also designed and built a number of special machines to turn out the work. Among these is a 6-foot boring and milling machine having a 48-inch milling cutter head. Mr. Dean worked in every department of the Indianapolis plant. He had a common school and high school education. Working in the shop in vacations from an early age, before he was out of school he had learned the machinist's trade. At eighteen he quit school and went to work in the foundry, learning this trade. He was specially instructed in all branches of operation, with the idea of eventually taking charge.

In setting out to record some of the results of his experience Mr. Dean planned to set forth in compact form what may be called the economics of shop operation; to indicate on what lines he had succeeded in increasing output in all departments and in reducing overhead expenses. The plant in which he has done his work employs about 200 men and is of the right size to enable the man in charge to learn all the practical details. It is no theory, therefore, that he presents, and extended discussion has been avoided, the purpose being to point out how and where money can be saved and the efficiency of the whole productive machine increased.

THE PUBLISHER

TABLE OF CONTENTS

ARTICLE I

ACHIEVING INDUSTRIAL SUCCESS

Seven Cardinal Points—Profit the Object of Every Business Enterprise—Management the Basis of Success — To-day's Opportunities Pre-eminent

PEOPLE say of a successful man, "He is always lucky" or "He hit it just right." Luck is not the cause of success. Success is due to management and nothing else. One of the largest wagon makers in this country started business at a time when other wagon firms were failing.

A wheel works, on the point of failing, was taken over by the bookkeeper and a foreman. The two men in time paid off all the debts, and made a large fortune and retired while still comparatively young. This was management—not luck. Rockefeller would have been successful no matter what business he might have chosen, because he had great managing ability.

The Seven Points of Success

To achieve success, seven factors must be kept in mind. Attention must be given to all of them. These factors are:

1.—*Publicity.* Keep your firm's name, its location and its product in every possible buyer's mind.

2.—*Selling Price.* Lower the selling price as business gets dull. Raise it as you get busy. To prevent selling too low fix the selling price at a point high enough to turn only one-quarter of your inquiries into orders.

3.—*Buying.* Save all you can in buying, both in price and in quantity. Keep as small a stock of material on hand as possible.

4.—*Employees and Assistants.* Surround yourself only with capable men.

5.—*Design.* Design the parts of your product so that the machine work and assembing will take the least possible time. Continue to change the design as long as improvement can be made.

6.—*Low Manufacturing Cost.* Never allow the time for making a piece to be greater than the shortest record time for this piece. If it takes longer, make the foremen explain why, and if there is a fault in the machine tool, jig, or material, correct the fault. This will bring profits.

7.—*Thorough Inspection.* Know absolutely that every machine shipped is perfect. This will give you a reputation for fine machinery, which is the best and cheapest advertisement you can have. Inspection of parts will save expense in assembling, and that will reduce your manufacturing costs.

Making a Fortune

The object of every enterprise is to make money—to make a fortune, if possible—for the owners. Luck plays a small part in its ultimate success or failure. Good management and the observance of all the seven points of success determine the final result.

Years ago our forefathers moved from rugged New England, with its fields full of boulders, into States farther west, where the virgin soil raised immense crops and made many people rich. The generation following looked back and said, "What opportunities there were then!" Each succeeding generation still looks back on the preceding one, and says the same thing.

To-day we look back at the prices secured for our product 30 years ago, and say, "What an opportunity there was for making money then." At that time the price of manufactured machinery was 50 to 100 per cent. higher than it is now, for the same machinery. As a matter of fact, the opportunity for making money at the present day is greater than it ever was. A firm manufacturing a competitive article can make a comfortable fortune in from 10 to 20 years, if it can sell to the full capacity of the plant and get the full output of the machines, tools and assemblers. The way in which it is possible to secure these desired results will be explained at more length in the articles dealing with machine shop operation and management.

ARTICLE II

DUTIES OF THE SUPERINTENDENT

How He Ought to Study Details of Department Operation — Graphical Production Records and Their Use to Secure Co-operation of Foremen

FULL authority should center in a works manager or superintendent. He should be a member of the firm. Many a company has come to grief because it did not have a practical man in the firm, but depended on hiring its practical knowledge.

The superintendent's office should be located at the most prominent point in the works, so that he will be very accessible to every one. "The master's eye enricheth the soil." If he locates himself off in the main office many important matters about the works that he should be informed of will never be brought before him. The men will say to themselves: "I will tell 'Super' the next time he comes around," but by that time the whole matter is entirely forgotten. The superintendent's office should be of glass and should be considerably raised so that he can look over the whole shop at a glance. Papers on his desk cannot be seen by any one standing outside if his office is raised.

What the Superintendent Ought and Ought Not to Do

The shop-material routing clerk should be located with or near the superintendent. The superintendent's duty is to increase the output per man with no reduction in the quality of the product. This is the superintendent's most important duty. I am afraid that we think of him as a busy man at his desk or out in the plant directing the work, asking how this is getting along, what is lacking here, etc.; pushing those jobs forward that the customers are in the greatest hurry for. This is what the office would like to have him do, and there is a temptation for him to do it. When he does so, though, he is out of his place; he has dropped to a mere routing clerk.

To study output per man and quality of product is the superintendent's most important work. If he does not spend three-fourths of his time on this subject he is not earning his salary.

The superintendent should be a faddist on cutting speeds and feeds and the number of cubic inches of metal removed per minute. He should keep in mind that at the point of the cutters and in the fixtures that cut down the idle moments between cutting operations lie the firm's profits.

He should be everlastingly correcting the allowance of finish on the patterns, so that the rule can be followed of taking two cuts and only two, for any one operation. Taking two cuts where three were previously permitted will cut one-third off the machining time.

He should be constantly studying the design of his company's product, changing it to eliminate labor in manufacturing. The brains of the concern should be kept on this work as much as possible.

How He Keeps Track of Production

A complete itemized set of reports, to be presented in these pages later, that give in detail the amount of work turned out by each department and by each class of men in each department, should be handed to the superintendent on the first of each quarter. These reports will enable him to catch immediately a dropping off of output per man in any department, and to correct the trouble. From these reports he will know which are his best foremen—the ones to be shoved ahead.

The superintendent should give all his orders in writing, keeping a copy of each for follow-up. He can never assume that his orders will be carried out without being followed up, nor must he try to depend on his memory for his following-up system. A good system is for him to have a spike file on his desk on which he sticks copies of all hurry-up orders. The first thing each morning he goes over these orders with his assistant, who tells him which orders have been carried out and which not, and the reason why not.

All notes that refer to orders that need not or cannot be attended to immediately he will file in one of three pigeon holes; one marked "First of Month," one marked "Fifteenth of Month," and one marked "First of the Quarter." Filing in these three places for future reference is a better way than filing in a daily tickler, as the notes are brought up only on certain days far apart. The superintendent thus keeps clear of this clerical work on all other days.

The Superintendent's Daily Routine

The superintendent must keep in mind that all clerical work that he attends to is dead loss to his firm. He must look at it in the light of a necessary evil. A very good system that will save him time is to carry around a folding partitioned pocket book with divisions in it that represent the various factory divisions. Into these compartments he drops the notes referring to work in these departments. Everything in the plant can then be attended to in one trip through the plant.

A suggested daily routine would be as follows:

1. Take all the notes for the day from the tickler or spike file.

2. Assort them, placing those in the pigeon holes or tickler that are for the future, and those for the day in the folding pocket book, to be taken around the plant.

3. Pass over the order sheets to the shipping department for those machines that are to be shipped on the day.

4. Take the folding pocket book through all departments. Rush up the work that is lagging back.

5. After this is done take up the desk work and the new original work.

Systematizing for the Daily Investigations

The superintendent's range of eyesight being limited to the small area of a few feet around him as he walks through the plant makes it necessary for him to depend on something more far-seeing than his eyesight to keep track of conditions in the plant. These department records give him this insight to all conditions. They are a continual watch over the whole plant.

Each day the assistant superintendent should ask the machine foreman:

1. What troubles are you having from faulty foundry work, or bad material?

2. What fixtures and appliances do you need to carry on the work in a better way?

3. On what castings could the amount of finish be reduced?

He should ask the assembling foreman:

1. What troubles are you having with the work from faulty workmanship or material?

2. What pattern changes or changes of design could be made to eliminate assembling?

3. What parts are lacking to complete the erection of machines now being assembled?

The answers to these questions should be given to the superintendent each day in order that he may take the matter up with the proper parties. If this is done each day it will make a material difference in the profits at the end of the year.

The assistant superintendent should keep copies of the record cost cards of all the parts of the popular-sized machines that the firm turns out. All new cost cards he will compare to the master cost cards. Any falling down will require an explanation from the foremen. Particular cases he will report to the superintendent.

Supplying Material and Tools to the Machine

To turn out work with the least loss of time it is important to have a full week's work ahead at each machine tool; otherwise, the rate of output will drop. It is impossible to handle the supply to the machines to such a nicety that one can get along with less. Crowd together the machines which work on small pieces. Leave plenty of space around the machines which do large work.

A number of large machines set around a space common to them all will require less total area than if set apart, each with its own separate storage floor.

There might be a card or order system at each machine which will tell the workman what his next job is and what is the job that follows it. He will then be able to reduce the idle time of the machine between jobs. He can plan his work, get his drawings, tools, straps, bolts, etc. If there is a fixture or jig he can get this from the toolroom, and he can get the piece itself ready to put into the machine. If there is anything wrong with the tools or jigs there will be time to have them repaired.

A good system is to have a row of hooks on a board at each machine and hang the orders on these hooks in the order that the jobs are to be done. The system can be elaborated. The assistant machine foreman can pass from machine to machine and hang different colored tags over the orders, signifying whether the tools and fixtures are all in condition for the job, or whether they need grinding or repairing. A white tag hung on an order would mean that the casting or piece is at the machine and the tools are all ready to carry out the work. A red tag would mean that the casting is at the machine but that the tools for the job are in the toolmaker's department being ground or repaired. A blue tag would mean that the casting or piece is at the machine, but the tools have not been inspected as yet to see that they are in fit condition to carry on the job. No tag hanging over the order slip would mean that the casting or piece has not yet arrived at the machine, although it is expected.

If such a system is carried out the workman can take the very best advantage of his work. He will be able to use some of his loafing time advantageously to the firm, and this will show at the end of the year in increased plant capacity and profits.

Information to be Kept in Graphical Form

The superintendent should use graphic diagrams for as many of his tables as he can. They present a picture of conditions in a most comprehensive manner where mere tabulated figures fail. These graphic tables should be in a loose-leaf book so that extensions can be made. In this book he should have the following, shown graphically. They should go back as far in the history of the firm as he can get the information:

1. *Pounds Shipped.* Pounds of finished machines shipped by the firm each quarter of a year.

2. *Total Hours.* Total hours put in by the workmen each quarter.

3. *Machine Hours.* Machine department hours quarterly.

4. *Erecting Hours.* Erecting department hours quarterly.

5. *Pattern Hours.* Pattern hours quarterly.

6. *Foundry Hours.* Total foundry hours quarterly.

7. *Non-Productive Hours.* All hours quarterly not taken care of in 3, 4, 5 and 6.

Divide the number of pounds in the first table by the hours in each of tables Nos. 2, 3, 4, 5, 6, 7. The quotients give the number of pounds of finished machines that each man's labor helps to turn out per hour. Multiplying this by the hours in the work day gives the pounds of finished machines manufactured for each man's day's labor. This gives the following graphic diagram tables:

Tables of Labor Production

8. *The Whole Plant.* The number of pounds of finished machines turned out by the plant daily, averaged per man.

9. *Machine Man.* The number of pounds of finished pieces each machine man turns out per day.

10. *Erector.* The number of pounds of finished machines each erector turns out per day.

11. *Pattern Maker.* The number of pounds of finished machines being made for each day's work of a pattern maker.

12. *Foundryman.* The number of pounds of finished machines manufactured for each day's labor in the foundry per man.

13. *Other Help (Non-Productive).* The number of pounds of finished machines manufactured per man for each day's labor of help other than already enumerated.

From tables Nos. 8, 9, 10, 11, 12 and 13, especially from Nos. 9, 10 and 13, the superintendent can see exactly what improvement each foreman is making in his department. The superintendent can commend the ones that have made progress and give a ginger talk to those whose showing is poor.

The same system of tables can be worked out on the basis of the proportion of the wages to the output, thus: 14. Quarterly pay-roll. 15. Machinemen's pay-roll quarterly. 16. Erecting men's pay-roll quarterly. 17. Pattern pay-roll quarterly. 18. Foundry pay-roll quarterly. 19. Quarterly pay-roll for the non-productive labor or all help not already covered.

Tables of Labor Cost of Product

Divide the pay-roll in each of the above groups of employees (Nos. 9 to 19 inclusive) by the number of pounds of finished machines shipped, No. 1, and multiply by 100. The result gives the pay-roll-cost in each department for 100 pounds of finished machines shipped. It is better to take 100 pounds rather than 1 pound as a basis for the tables because it gives a larger figure for the result, and any small change from quarter to quarter will be more noticeable. This gives us the following tables:

20. Total plant labor cost for each 100 pounds of finished machines.

21. Machine labor cost for each 100 pounds of finished machines.

22. Erecting labor cost for each 100 pounds of finished machines.

23. Pattern labor cost on each 100 pounds of finished machines.

24. Foundry labor cost for each 100 pounds of finished machines.

25. Non-productive cost on each 100 pounds of finished machines.

Watch the hours in relation to the tonnage output rather than the wage relation to the tonnage. Thinking too much about the wage side of the business will end in not giving men raises at the proper time and in losing the best men and keeping good-for-nothings. High priced men are by far the cheapest. In rare cases underpaid men will work for the interest of the firm, but generally they more than get even by shirking. Decrease the hours and increase the tonnage and the wage cost will take care of itself.

Tables of Foundry Operations

The separate divisions of the foundry should be watched by tables based on monthly records instead of quarterly ones, as follows:

26. Total number of pounds of castings reported made by the foundry.

27. Number of pounds of castings thrown out by the machine shop due to bad foundry work. These are to be deducted from the castings in table No. 26 and give table No. 28.

28. Number of pounds of good castings made by the foundry.

29. Number of pounds of iron charged into the cupola.

30. Number of pounds of scrap from the foundry (sprues, risers and castings that turned out bad in the foundry). The figures had better be gotten by a process of elimination rather than by direct weighing. Take from the total scrap charged into the cupola the amount of scrap obtained from the machine shop plus the bought scrap. This gives the scrap that came from the foundry.

31. Number of pounds of castings the foundry should have reported as made, barring the loss of iron in the cupola, the dump and shot iron in the foundry and cleaning room, the figures being obtained by subtracting the figures of No. 30 from those of No. 29.

This cycle of foundry weight tables can be used as a check on the weighing and is valuable in keeping the weigh men accurate. If they know that you have a rough check on them they will be careful. The following men are checked: The man who weighs bought scrap iron; the man who weighs castings thrown back from the machine shop; the men who weigh the castings made by the foundry, and the man who weighs the charges for the cupola. I know of a case where this system of checking discovered the crookedness of a foundry foreman who was sending in reports of weights of castings made by the foundry higher than the actual amount so as to make the casting cost appear low.

Further subdivisions should be made as follows:

32. Total foundry hours.

33. Total molders' hours.

34. Hours foundry helping.

35. Hours casting cleaning.

36. Hours coreroom.

37. Hours of other foundry help (crane men, cupola men, foremen, etc.).

Dividing the weight in table No. 28 by the number of hours in tables Nos. 32 to 37, inclusive, and multiplying by the number of hours in the working day gives, the number of pounds of castings worked on or helped on for each day for each man in any of the groups of men in the foundry. This gives the following tables:

38. The day's output of each molder.

39. The number of pounds of castings each foundry helper helps on each day.

40. The number of pounds of castings each casting cleaner cleans each day.

41. The number of pounds of castings a day each man in the coreroom makes cores for. In this table the core-

room helpers and foremen are not separated from the core makers as it would be a useless refinement.

42. The number of pounds of castings turned out per day per capita of all other help (cupola, crane and foundry foremen).

The same system of tables can be worked out on the basis of wages, thus:

43. Total foundry pay-roll, including in this the pay-roll of No. 48.

44. Total molders' pay-roll.

45. Helpers' pay-roll.

46. Coreroom pay-roll.

47. Crane, cupola and foremen's pay-roll.

48. Pay-roll spent in other departments for doing foundry repair work.

Dividing the pay-roll amounts in Nos. 43 to 48, inclusive, by the weights of good castings turned out (No. 28) and multiplying by 100 gives the labor cost on 100 pounds of good castings made by each man each day in any one group of men. This gives the following:

49. Total labor cost on 100 pound castings.

50. Molding cost per 100 pound castings.

51. Helping cost per 100 pound castings.

52. Core cost per 100 pound castings.

53. Crane cupola and foremen labor cost per 100 pound castings.

54. Foundry repair labor cost for each 100 pounds of castings.

These graphic tables are not so much work to take care of as would seem at a glance, for with the exception of the foundry they come up quarterly only. This information is laid on the superintendents desk by the bookkeeping and cost-keeping departments in sheet form.

Down the left-hand edge of each of these sheets are given the names of the items and the amounts are placed in columns at the right. The sheets have spaces for each of the four quarters of the year, for the total year and for the previous year, the items being reduced to the quarterly rate, so that comparisons can readily be made.

The superintendent should transcribe these items to his graphic-diagram loose-leaf book with his own hand. This will bring the various changes of conditions throughout the plant home to him in a much stronger way than if his assistant transcribed them. From these graphic reports the superintendent can catch the slipping backward of any department before it has gone too far.

A foreman who is a genius will improve his record in the face of a change in the class of work that requires increased labor per pound of product. The graphic table line will continue to climb.

It is a splendid idea to give each foreman a quarterly written statement of the number of pounds of finished work his department averages per work day per man. This acts as a spur to the foremen. The foremen take the keenest interest in these reports. To them the reports are a tangible statement which they can compare with their records of previous years. These reports keep reminding them that any hours wasted will injure their record. A lazy man, a man who spoils work, or a slow man will not be tolerated, because such men increase hours without increasing the pounds output.

The foremen will call the superintendent's attention to faults that hold the work back, such as ill-shaped castings, requiring hand chipping; too much finish allowance by the pattern shop, obliging the machine department to take three cuts where two would have been enough, or inaccurate work from preceding departments which consumes time in correction. I remember a machine foreman's remark when forced to throw away a casting that had turned out bad after machining. "Look at my hours being thrown into the scrap box!" One foreman was so interested in his report that he always carried it around under his work cap.

Department Watching

Once every two years, or if he can see his way clear to do it once every year, the superintendent should spend a full month in each department. He should get there every day at whistle time in the morning and stay in the department until the close in the evening, criticising and making improvements. He should let everything else drop and take up one department at a time in this way.

Of course the average superintendent will say: "Oh, this is absolutely impossible. I have my regular duties to perform each day." What are these regular duties? They are all mere clerical duties which he should not be doing at all.

The superintendent should know how to use an assistant and should throw all simple work and statistical work upon him. This will leave the superintendent free to do original work and improvement work. The assistant must have an absolutely accurate mind. He must be fearless in pressing the work forward that the superintendent wishes done.

His next question will be, "What will become of the work in the other departments if I never go into them for a month?" He can take them up some following month and really do some good in them. The improvement that the superintendent makes in a department by walking through it a few times a day, superficially looking it over, is practically nothing. He can tell nothing about the detail way they run except by staying in the department all day, each day, for a month. The things that he will straighten out by doing this will make an enormous profit at the end of the year.

Remember the average machine shop only turns out 30 per cent. of its possible capacity. This leaves 70 per cent. to be worked on. I know a plant where the superintendent's adopting this system increased the output per man and reduced expenses in all departments. In the foundry alone the cost of casting dropped from 3 cents to $2\frac{1}{4}$ cents per pound. This meant that he sacrificed his $2.50 per day clerical duties and his occasional walk through the different departments which were always on dress parade during the time he was in the department.

ARTICLE III

SELECTION AND HANDLING OF MEN

Hints Looking to the Development of the Able and Contented Working Force Sorting the Men According to Ability — Shop Thieving

SURROUND yourself with capable men. It means more profit to your firm than any other item of the business. There is every argument in favor of employing only the best talent, and no argument against it. If you can get good enough men, you can afford to turn them out millionaires, as Marshall Field did his partners: he made millions doing it. Both Andrew Carnegie and John D. Rockefeller attribute their success to the fact that they surrounded themselves with very capable men. They have been preaching this to the American business man ever since. Surround yourself with the same kind of men they did.

Qualities in Department Heads

Don't be satisfied with the department heads until you get men that manage their departments better than you could yourself. With capable men around you, the work will be done easily, will be turned out cheaper and better, and the output of the factory will be increased. It will give you time to take up the big problems that spell profits. It is team work that builds up an establishment; you can't do all the work.

It is better to have a fine foreman even if he stays with you only a few years and gives you the trouble of breaking in a new one than to put up with a poor one just because you know he will stick with you forever. No matter what other quality a foreman lacks, he must not lack energy. A lazy foreman with brains will bring poorer results than a medium-brained foreman who has unlimited energy. The energetic man cannot put up with loafing or laziness; he does not understand it. The lazy foreman, in spite of himself, cannot help sympathizing with laziness. He knows the disease.

Giving Free Rein to the Capable Man

When you do get the exceptional good department head, gain his perfect confidence. Give him full sway, it is the only way to get the most from him. If you can see your way clear, never turn down any of his proposals. Let him carry his ideas out even though some seem impracticable. Giving a good man rope will stimulate him and bring great results. If you continually turn him down, he will give up proposing things. He will get out of the habit of scheming for better things. A very good thing may occur to him which he will not speak to you about because he is sure you will turn it down.

Be sure to have understudies coming along for every responsible and semi-responsible position. You are then perfectly guarded. Your work will always go smoothly.

Good workmen are not expensive when the value of their output is compared to the wages paid. The dollar paid in salary to the best man in the shop brings in a larger return to the firm than the dollar paid to the poor man. It is quite common to see the best man turn out double the output of the poor man, yet the best man will not receive double the wages that the poor man gets. There is about one good man in six. It certainly pays to worry through, trying the bad five, to get the sixth.

Fitting the Man to the Work

Don't try to make a man over into something that he is incapable of being. A man cannot change himself, as far as his character goes, more than 10 per cent.

Put on the physically and mentally fit man for each job. For very heavy work, where the output depends largely on the physical effort put forth by the workman, you must get a powerful man who has great endurance. The person with only average strength does not realize what powerful men there are in the world. They are by no means rare, either.

Continue trying new men until you surround yourself with geniuses in the line of work you want them to do. Don't be afraid of greenness. Greenness is the least of the faults in a man; it rapidly disappears.

Taking Men to Task for Mistakes

Never call down a man for breaking a machine, no matter how expensive the break may be, if it was done by forcing the work too hard. Better tell him that you are proud of him. He was showing the right spirit. Few men push the work to the limit. It is wrong to take the man to task who is really a rare article, a gem, because he was overzealous. This, of course, does not apply to the heavy-handed, careless man who breaks his machine by dropping something into the gears, or throws on the wrong feed by mistake. Even this man, if he be a big producer, should be dealt lightly with. If he is an habitual blunderer, get rid of him; he has an inaccurate mind.

Never continue to scold a man after he gets mad. Walk away from him until he cools down. Many a good employee has been lost by not knowing when to walk away

After you give a man a raking down for something he has done, go back and brace him up with cheerful words. When you do this, he knows that your criticism was of his work, not of his ability. He knows that you are satisfied with him as a man; the criticism was purely of the work.

I have always found that more can be done by praise than by blame. Persuasive talk accomplishes much. Men who require a blowing up in order to keep them to the mark had better be dropped. Of course there is a difference in people; all cannot be handled alike. It is a good thing to give the foreman and sub-foreman a little talk, once in a while, such as "Let's all pitch in and pull together to make a fine showing. The selling department claims to be beating the manufacturing department." Make them feel that it is a game in which to win, each one must push the work to the utmost.

Helping Employees in Trouble

Always help the workman when he is in trouble. When he gets hurt give him half pay. At such a time he needs the money desperately, and the outlay is but a slight item in the general expenses.

If he asks your advice on a legal matter, or wants your advice about an investment, or a medical matter, give him all this advice you can.

Close the plant half a day for the funeral of your old employees. Go to the funeral yourself. You must be a sort of father to the workmen. Try to eliminate the soulless corporation feeling.

Lifting the Esprit de Corps

Always hire young men. Never lay men off for old age. Change them over to watchmen, gate watchmen, sweepers and roustabouts on light work, at reduced pay. Use them for picking up odds and ends of material.

They make ideal men for these positions and earn their wage. The trouble with most firms is they hire old men for these positions, instead of taking their own old men.

By getting rid of one or two men who are disturbing elements, the output of the whole shop can be increased. One or two men in a shop make a practice of telling the others that they wouldn't do this, and wouldn't do that —"don't be a horse," etc. These very men do not practice what they preach at all, but do good day's work. Their fellow workmen are blind to this hypocrisy.

Often foremen and workmen think they are turning out the maximum of work when they are not. Send the foremen to observe the work at other shops. Bring on machine demonstrators from the machine factories. They are sent free of expense, generally. Borrow a rapid workman from some other shop on a holiday. Let him put up a day's work. This will produce an enlightening effect.

Buying Work from the Workman

The amount of work a man should do must be looked at in this light: The employer is buying work from the workman. Like other commodities, work worth a dollar should be obtained for every dollar spent. If the employer pays for 10 hours' work, he is entitled to the full 10 hours' work. This means the workman must start to work at whistle time—not begin to get ready to start then—and must not quit until the quitting whistle has blown.

The sole and only object of manufacturing is to make money. This should be kept continually in mind. A very large output per man is essential to success. Everything else is dwarfed in comparison to this. The so-called humane person may say that this is a hard way to look at the subject of employing our fellow men. A firm that looks at the subject differently fails. Failure generally brings misery to all concerned, including stockholders. Failure forces great hardship on the workmen for no fault of their own. Failure generally happens in hard times, when work cannot be easily found. Failure is bad for the old workmen. They may never be able to get good jobs again. They are changed from good, prosperous men, possibly with bank accounts, to men with all their life's saving gone. Prosperous firms are able to stand the strain of hard times, and their workmen share their prosperity.

Employer's Acts Having an Adverse Influence

Do not drive up to the plant in your $3000 automobile. The workmen will think it cost you $5000. They cannot help comparing the price of your automobile with the price of the home they have been trying to buy all their lives.

Do not put in window flower boxes and do not take employees to baseball games. In other words, do not do anything that looks as though you were throwing away your money. I never yet have seen a case where it did not bring out the remark from the employees: "Well, they must be making lots of money. They ought to put that into our pay envelope."

A Quarterly Method of Sorting Employees

Every quarter get from each foreman a list of the men arranged according to their value or future value in a department. Tell the foreman that you want to know who, of all his men, he would hate the most to have quit. Then, who would be the next, etc., etc. Tell him not to place them in the list according to their wages, but according to their value. Ask him which men seem to be capable of rising more rapidly than the others. They are the ones that will be valuable to you in the long run.

Those who cannot rise, replace with green men. This system will surround you with producing geniuses.

This once-a-quarter talk has a tendency to keep constantly before the foreman the fact that he has men of different calibers. It puts it clear in his mind so that he will get rid of his poor men and hang to his good men.

Training One's Own Skilled Help

Make your own mechanics. Weed out the poor ones. It is the only possible way to get the valuable men. You cannot hire them from other shops. The good workman rarely shifts from one shop to another, because his employer, appreciating him, will not let him go. The mechanics that you will teach will do the work your way. They will stay with you as they are not sure they could hold jobs outside.

Hire bright laborers and teach them the trades. Take men, not boys, for the man who has had to struggle along at laborer's wages, supporting a family, will do the most good and be satisfied with his pay if you treat him fairly. The opportunities for laborers to become skilled mechanics are so few that your efforts will be appreciated. The laborer will be satisfied with a more moderate rate of wages than the shifting mechanic that you hire from the outside. At first your foremen will hate to be teachers, but soon they will get used to it and prefer it.

It is best to have a backbone of speedy, accurate, all-around high-priced mechanics through your plant. This will be a skeleton framework around which the force is built.

If the system of breaking in green men, instead of hiring outside skilled mechanics, is followed, a plant will never be troubled about scarcity of labor even when other shops in the locality cannot get men. It requires time and patience and a free hand at laying off the unsuitable ones. Only about one man in six will be found suitable to learn the trade, but the trouble of trying them pays many times over in a large output per man and a large output per dollar spent in pay roll, and this is the real object of the game.

A contented family feeling springs up in a plant managed in this way, which is a fine thing. Labor troubles are eliminated. The paid agitator never poisons the minds of the men, because he is never hired under this system. The men appreciate what has been done for them and value their positions. The plant gets a good reputation among workmen and draws the best talent from the locality.

The Employment of Boys

Use boys where it is practicable. The best 35-cent per hour man cannot compete with a good 14-cent boy on plain work. To come out even, the man would have to turn out two and a half times the boy's output. Here is an actual comparison from the cost cards of cost of work when done by boys and by men. The job was machining and drilling 100 cylinder heads by two men getting 21 and 24 cents per hour respectively and by two boys getting 7½ cents and 5 cents. It took the men 12 hours to get out 100; cost $2.74. It took 16 hours for the boys to get out 100; cost 97 cents. This card was made out a number of years ago. Both the men's wage rates and the boys' rates were a good deal lower than they are to-day, but the ratio will run about the same.

Only a small proportion of the boys will stay with a firm longer than 4 years. They feel that they must make a try in the world outside the shop where they learned their trade. I have found that the proportion of boys that can be taught trades runs much higher than the proportion of men that can be taught; that is, less weeding is necessary with boys than with men. The reason for this is that there has been very little picking over of the boys, so that there is a good chance of getting a boy who will represent the average of the whole boy kingdom. The men who are out of work are likely to be the poor leavings that no one could use, so the chance of getting a genius is not so good as with the boys.

Shop Thieving

Thieving is generally done by recently employed men. One gets acquainted with his old employees; they become part of the family. The transient is the one who steals. Bar off and lock up the department of the plant where valuable material is kept.

In order to stop thieving, the superintendent must be on such terms of familiarity with his men that they feel free to inform him of irregularities. This is his grapevine telegraph through which he gets his inside information. The manager or superintendent has but two eyes. As he walks through the plant, his field of vision, as already stated, is limited to an area of a few square feet. This area is on dress parade. The things that go on when he is away are what he should know about. The good men need no watching. They are by far the great majority, otherwise our social system would not hold together.

No system of checking or red tape will discover stealing. I remember a case in a factory in our town where stealing was going on. I happened to know, personally, a young man in this factory who told me the incident that led to discovering the stealing. One day the wife of one of the employees came into the office. It seems she had been having a row with her husband and he had left her. She said, "I don't think it is right the way my husband is stealing from you." This was the first they knew of any stealing. An examination of his barn disclosed an enormous mass of stolen material, small tools, etc., including a 3-hp. electric motor bought of outside parties to drive a small new lathe. This company had a wonderful system of checking, and yet the theft of the motor was not discovered by it.

I know a case where a man continually stole the sweeper's broom. The broom was marked with the firm's name, and a large cross was painted on the straw part. Only a very shortsighted man could suppose that the broom would not be missed immediately. This man stole four or five brooms before he was caught. I also heard of a man in a big electric company just outside Pittsburgh that had a sack made in the lining of the back of his coat. He came in the morning a straight man, and went out round-shouldered, from a mass of sheet brass clippings under his coat. He tried to steal the intershop telephone. This is where his shortsightedness came in, and he was caught.

Courts give light sentences to thieves, on account of the sympathy of the public, and thieving is profitable though the thief may serve a prison sentence occasionally. Thieving is a serious loss to both employers and employees, and requires constant watching. Thieves are very lacking mentally, and are easily caught by being led into a trap. For example, a number of pieces of brass can be left around where the thief will notice them and will steal them. If this is repeated, with some one watching the pieces, he is easily caught. After making sure of the thief dismiss him. The main point is to get the thief out of the plant. It is a duty to the honest workmen, as well as to the employer.

Increasing Wages as Time of Service Increases

Start the green men whom you expect to make into mechanics at the standard rate of laborer's wages paid in your locality. Make up a table of raises that will retain the men in your service. Give the men raises in accordance with their length of service. Give the increase by half cents. Don't wait until it is time to give a man one cent, or two and a half cent raise. The little constant sugaring of the pay envelope keeps them encouraged, and makes them realize that their work is appreciated.

It may seem an unjust system to give raises according to time of service, and not according to the relative value of men, but it is not. If you get rid of poor learners, you are sure of a force of men somewhat even in their mental capacity, and other things being equal, the longer a man stays with you, the more valuable he becomes.

Besides this, it is a very satisfactory system for the man. He comes into your employ, we will say, a young man, just married, with low expenses. He and his wife are boarding possibly with her parents. His expenses gradually increase. A baby comes into the family. His wages are going up to keep pace with his increase in expenses. He rents a house; more money is in the pay envelope. He has more children; again a raise. After he has been with you 6 or 8 years, the old folks come and live with him; more expense. By that time, he is making good pay. See what a satisfactory life he is leading. He will continue to get raises as long as he stays.

Now, instead of this, suppose you had jumped his wages to the full amount in three years' service. He would have jumped up his expenses. In other words, the fixed rate of raises fits the natural fixed rate of expenses incurred by a man as he goes through life. There is nothing an old employee hates so much as to see newcomers paid more money than he receives.

The Wages Increase System in Its Working

According to the above-mentioned table, a man's wages increase rapidly during his first year, less rapidly during his second year, and still less rapidly during the following years until he becomes an experienced workman, when his rate is fixed in comparison to his standing with the other men. Those who are not satisfied with the table rate of increase, should be allowed to leave. The man who expects mechanic's full wages before he has learned the trade is not a desirable employee. A dissatisfied workman is unprofitable, and creates dissatisfaction among his fellow workmen.

On hiring the green man tell him just what his raises will be and when they will come. He will take more interest in the work knowing there is a great future for him.

When it comes to giving a man a raise let the foreman notify the man. When it comes to the refusal of a raise asked for, let the superintendent notify the man. This will keep a good feeling between the foreman and his men. I do not mean by good feeling the catering of a foreman to his men. A foreman of that sort should be gotten rid of immediately, as the men will run him to death, and you probably will wind up with a strike in his department.

Another good system of raising wages, which will apply to a small department—say the coreroom of a foundry—is to pay the best workman a rate higher than the standard for his trade. Pay the second best man the standard rate, the third man under the standard, the fourth still less than the third, etc. The newest learner will get a rate about 25 per cent. greater than laborer's wages.

ARTICLE IV

PIECE WORK AND BONUS SYSTEM

Conditions Favorable and Unfavorable to Piece Work — What to Consider in Establishing Rates — Observations on the Physically Fit Man

PIECE work or the premium system is applicable to any work that can be thoroughly inspected. It is not suited to the assembling of fine work, but is all right for the assembling of the rougher classes of machinery where the designer has allowed ample leeway for poor workmanship, where the customer pays a low price, and where a perfect machine is not absolutely essential. The output per man in a plant changing from day work to piece or premium work will increase in the ratio of three to five.

When Piece Work is Not Advisable

Piece work or premium work can be used on the machine operations of the automobile engine. All pieces can be gauged and inspected as they come from the machines. Piece or premium work cannot be used without danger of a bad product on the assembling work, because thorough inspection here is impossible. An assembler can screw a stud carefully into the hole that has been partially stripped and the job will pass the inspection and the running test. When the buyer tries to tighten down the nut on this stud, on account of a leak, the thread will be stripped.

An assembler discovering that he has not placed enough liners or shims between the connecting rod and the cap to prevent the pinching of the crank, may leave these bolts loose rather than waste his premium time in correcting the matter. The engine will pass all tests, but give trouble afterward.

The ground bearing of the valve on its seat may be broad on one side and narrow on the other and still be tight and pass the test. This valve will not stay tight as long a time as one that has an even bearing all around.

Inspection will not discover faults in assembling. The workman must be depended on to do good work.

Eliminating False Moves

To get the most out of piece work, both for the workman and the firm—for the workman a higher total wage, and for the firm a lower piece rate—every false move, no matter how seemingly insignificant, must be absolutely eliminated. This can be done to some extent by the workman, but can be accomplished better by having an intelligent overseer stand by the man while at work and call the man's attention to each one of his false moves. A good way is to have the man count aloud the number of movements he makes. He will soon be interested in the possibility of reducing this number.

The molder in the foundry making small molds, may be putting one too many shovelfuls of sand on his mold which afterward has to be struck off as superfluous. In striking off the mold he may make two moves where one long sweeping move would do the work.

Setting the Time on a Piece

When a man is able to do the task without making any false moves, time him. See that he is moving rapidly, that is, not holding back because you are timing him. From his time on one piece figure what his output would be for a day. The actual output will drop below this on account of small delays now and then. The allowance will be different on different classes of work. On work requiring very little physical effort, the allowance will be small. On work taking great physical effort, this allowance should be as high as 40 per cent. That is, a man will have to rest 40 per çent. of the time on the heaviest work. His resting generally takes the form of working slowly in the afternoon when he is tired, and quitting work rather early.

For instance, set the molding rates as follows on small molding after getting the true time of one mold when the molder is hurrying: $2\frac{1}{2}$ minute mold add 40 per cent.; total time $3\frac{1}{2}$ minutes; 5 minute mold add 40 per cent.; total time 7 minutes; $7\frac{1}{2}$ minute mold add 40 per cent; total time $10\frac{1}{2}$ minutes; 10 minute mold add 40 per cent.; total time 14 minutes.

On heavy piece work or premium tasks be sure to use a powerful man—a man physically fit for the work. There are men who never tire. The only effect enormous, continued physical effort has on them is to make them ravenously hungry. The energy expended is taken from the food the man eats and not from the man's tissue. He is burning food, not flesh.

I remember asking a workman who was doing very heavy work all day if he felt tired at night. He was a muscular, short, thick set man. He said, "No, I feel just as fresh at night after I have eaten my supper as when I started in the morning." He was physically fit for the task and felt no injurious effect from overwork. Few realize what an enormous amount of work the physically fit man can do.

A fireman on a big locomotive puts 15 tons of coal through an 18-inch door upon the fire in a run of five or six hours. At the Lake Erie docks men are paid 18 cents per ton for cleaning up the ore in the hold of an ore boat after the automatic unloader has handled all that it can (80 per cent. of the cargo). They have made as high as $12 per day of ten hours, which means 6.67 tons of ore were shoveled in one hour.

On straight work, not cleaning up, they are paid 13 cents per ton. When eight men are in a hold shoveling into 1-ton buckets each man handles five or six tons of ore per hour. A rate of eight tons per hour has been reached. The daily wages run as high as $6.50 to $7.80 per day. These are instances of the ability of the physically fit man for heavy work.

When skilled mechanics, in any trade, are paid $3.25 for ten hours, the piece worker will earn $4 to $4.50 a day; the exceptional man will be able to make about $5. These figures are for work where the piece rate has been correctly set and a good speed kept up all day.

Handling Work by the Specialist System

Where work can be specialized by having a man do but one or two operations, costs can be greatly reduced. This system of specialization increases the output per man and improves the quality of the product because each specialist is an expert on his one particular part of the work. A product made exactly to the drawings will result because interchangeability is a necessity to the system.

A firm can easily increase its force of skilled workmen, even when labor is scarce, for it is easy to break in green men who can be taught to do one or two operations only. Never make the mistake of putting a skilled mechanic on this simple work; it will be distasteful to him, and he will not be successful at it.

Automobile Engine Assembling

1st Gang—(Crank case). Stud crank case and ream bearings for crank.

2nd Gang—(Connecting rods and crank). Put in babbitt bearing and the bronze bush. Scrape bearing to fit crank. Number crank and connecting rods to keep them together.

3rd Gang—(Cylinders, valves, manifold and exhaust headers). Grind in exhaust valves. Fit on manifolds and exhaust headers.

4th Gang—(Wrist pin and piston). Fit wrist pin into piston, but do not fit in piston rings.

5th Gang—(Crank case and crank). Receive crank case from 1st gang and crank from 2nd gang, and fit crank into crank case, scraping bearings.

6th Gang—(Crank case and connecting rod). Receive crank case from 5th gang, and connecting rods from 2nd gang, and put connecting rods onto crank.

7th Gang—Receive crank case, etc., and put on gears, but not the cam shafts.

8th Gang—Receive the cylinders from the 3rd gang and the pistons from the 4th gang and the crank case from the 7th gang, and fit the rings to the pistons. Fit the pistons and rings to the cylinders. Fit the wrist pin to the connecting rod and bolt the engine together. The engine is now complete except cam shaft, etc. From here on it is transported on small chain hoist trolleys. In some places, on small wagons. Before this point the parts were handled by hand.

9th Gang.—Put in cam shaft, etc. Time engine and set valves. From here it goes to the belt test and regular test.

The same system is used on small parts, such as cam shafts, etc. Have a gang of all around mechanics to fill in when men are off and attend to troubles that are out of the ordinary. A part that requires excessive hand work, on account of bad machine work, is turned over to the all around mechanic, which prevents any stoppage in the natural flow of parts through the gangs.

ARTICLE V

RESULTS OF PRODUCTION METHODS

Methods Evolved to Minimize Clerical Work in an Establishment Having Foundry and Machine Shop Operations—Effect of Too Much System

MONEY is made at the points of the cutting tools and not at the points of the lead pencils. There is nothing quite so sad as a factory that has a clerical force out of proportion to the manufacturing force. Each producer must work the harder to support the parasite.

Beware of too much red tape. Avoid safe-guarding systems carried too far into details. Over-intricate cost-keeping systems do not pay as well as approximate systems. Many offices have shelves filled with time cards that cost hundreds of dollars, but which are not bringing in one cent's worth of returns. Putting this money into machine tools, or into efficient managers and high class workmen would be more profitable.

Too much red tape will tie the hands of the foremen and the employees. Many firms have failed because an intricate system of safeguarding killed the freedom of action of the foremen. Different foremen arrive at success by using entirely different methods. It is native ability that makes the thing go, and any check on this is damaging.

What Too Much System Did

In a certain plant an intricate shop system and cost-keeping system had been installed. The cards were most elaborate, and were printed in colors. All the foremen, even down to the straw bosses spent a large part of their time at improvised desks, shuffling time slips and writing. At a large machine the writer timed the cutting speed. The machine was running at one-third of its correct speed with a feed about one-half of what it should have been. The product of that machine, in fact, of the whole plant, was being turned out at one-sixth the correct speed. At the toolroom a crowd of men was killing time, likewise at the drinking fountain, and the tool grinding emery wheel. The only department in the plant that was running at all well was the foundry.

The foundry foreman was asked how it happened that he wasn't writing at a desk. He said, "When I took this job they handed me a book of instructions, explaining their systems. I took it home and studied it during all my spare time for two weeks and then gave it up as being too complicated to be practical. I turned it over to my assistant, and he and a high school graduate we hired tend to all the writing for the whole foundry. All I do is to have the assistant give me a sheet of paper each evening telling me how many castings are wanted on each pattern. I chalk the number on the pattern. It takes me about five minutes each day and I am then free the rest of the day to push the work through."

The concern afterward failed. It would have been better for it to have had no cost system, and to have set its selling prices at its competitors' prices and made the formen increase the output of the men until the firm began to make money. The "golden mean" lies between the two extremes of too much system and not enough system.

Conditions Which Demand a Routing System

As a plant grows a time comes when the foremen will not be able to keep everything in their heads. A system then will have to be installed for keeping track of every piece going through the plant. Lack of some of the parts for a machine being built is one of the greatest drawbacks to the assembling department. The machine cannot be completely assembled until the last piece is received and all hand work is finished on it. Workmen, of necessity, loaf on the job when all the parts are not ready for them.

The routing of parts through the plant will have to be taken care of by some one who is given authority. It cannot be left to department foremen. The foremen's whole attention should be concentrated upon improving the methods and the workmanship and increasing the output per man of the shop.

Previous to installing the following system in a plant employing 200 men, the foremen kept track of everything in their heads. They developed wonderful memories. Finally the amount of work grew beyond the capacity of their memories. Parts that were started through the works were lost. Duplicate parts were made. The finished stock room gradually filled up with special odds and ends that could not be used. Eventually they were scrapped at great loss. Often the failure to order a certain piece on a machine was not discovered until the machine was about erected. The smooth, regular flow of the work through the machines had to be stopped to push this piece through. In the meantime the erector loafed on the job until he received the lacking piece. Then the system described below was introduced:

System for Cost Keeping and Shop Routing

Two important points in this system are: (1) The shop part of the cost-keeping and the shop routing of material are in one system and in charge of one clerk. (2) The system is simple and requires but a small amount of clerical work. The writing consists of the mere jotting down of a symbol, or the drawing of a line through a printed word, or putting a cross or other mark in a space. This allows the foremen and workmen to give their full time to production. The system keeps absolute track of every piece that is going through the shop, and never lets one piece lag behind the others.

When an order for a machine is received, a number of acknowledgment sheets (Fig. 1) are typewritten in the office. One copy goes immediately to the customer as the acknowledgment of the order. A second copy is filed in the office to be used as the invoice to the customer when his machine is shipped. The office can always ascertain whether or not a customer's machine has been shipped by looking in this file. If the invoice sheet is there, the machine has not gone. If the sheet is not there, the machine has been shipped. A third copy of the acknowledgment goes to the drafting room; a fourth copy to the material routing clerk; the original sheet is sent to the superintendent.

The departments receiving these sheets file them alphabetically under the name of the customer. From this file the order can always be referred to if the customer's name is known. The office gives each order a consecutive number, known as the order number, which is eventually stamped on the finished machine to identify it.

During the process of manufacture all orders for material, all machine orders, and the assembling order for this machine, are marked with this order number.

Besides making out these acknowledgment sheets the office also makes out a set of sheets called the order sheets (Fig. 2). With these sheets is made out a card which is a facsimile of the order sheets and is called the erector's tag. Two order sheets go to the drafting

DEAN BROTHERS' STEAM PUMP WORKS

We Have Received Your Order for the Pumping Machinery Stated Below and for Shipment as Follows:

Indianapolis, Ind.

Ordered By

Ship To

Order Received Ship Via Your Order No.

Shipping Date Our Invoice No. Your Requisition No.

Quantity	Size	Style	Service	Our Proposition

ACKNOWLEDGMENT

Fig. 1—Sheet Acknowledging Order, Five Copies in All Used

room. The tag and one order sheet go to the material routing clerk, and the original lead pencil copy goes to the superintendent.

These order sheets are filed either by the order number or are kept in files that are sub-divided to represent the different departments of the plant. As the work proceeds the order is moved from one departmental division of the file to another. The sheet is always kept under the department that is holding the work back the most. For instance, it may be filed under drafting room if the need of a new drawing is holding the work

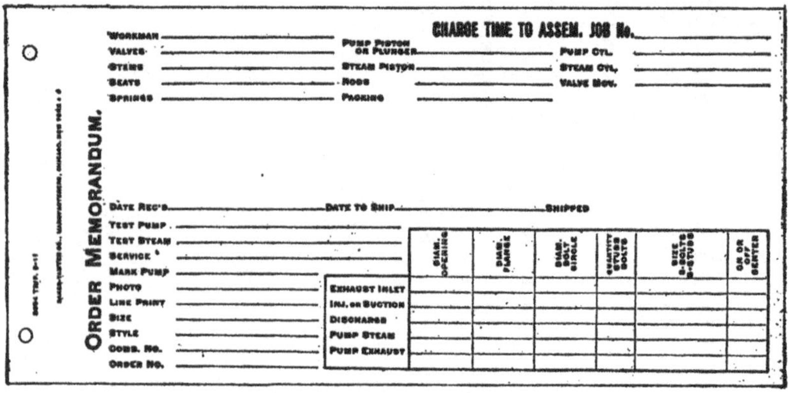

Fig. 2—Order Sheet. Size, 8¾x4⅛ In.

on the order back, or it may be filed under the pattern shop if new patterns must be made.

The orders thus filed save considerable traveling from one department to another. By taking the file of orders into a department, all the work can be rushed that the department is holding back. One trip a day, or even one trip every other day, is sufficient to cover everything.

The order sheet gives complete information to any one familiar with the product of the firm. It has written on it: The date of the order; shipping date; the name and address of the firm to receive the machine; the consecutive order number; the combination number; the size and style of the machine; all the peculiar or particular notes necessary for getting out the machine.

Each machine as it is designed is given a design number, which is called the combination number. This starts with No. 1 and continues indefinitely.

Every time any change in design is made in a machine, even if it be only a change on the smallest piece on the machine, a new combination number is given to the machine. For instance suppose that machine combination No. 15 should develop a certain weakness in some small part, say a little pin, and suppose the highest or last combination number was 800. This machine after the pin was changed would be given combination number 801, and would be built under 801 combination number until another change in design was made, when it would take on another combination number.

This is a fine system as it simplifies getting out repairs. A certain man does not have to be kept in the employ of a firm because he happens to remember how all the old machines were built. Another great thing about the system is the management is perfectly free to make all the improvements and changes in the design it pleases without in the least mixing anything or giving any further trouble.

The set of drawings for each size of machine is numbered with the combination number of the machine. Each drawing in the set is given a card number to separate it from the other drawings in the set or combination. Thus there would be for machine combination number 15, one set of drawings all marked "Combination No. 15" with the card numbers, if there were ten drawings in the lot running from one through ten.

After the change of the pin on combination No. 15, this machine would be called combination No. 801. All the combination 15 drawings would be used with the exception of the pin drawing No. 801, card 1.

Bill of Material

The drafting room makes out a master bill of material for each design or combination of a machine, which is a complete record of everything on the machine. It gives a list, in columnar form, of each piece used on the machine. At the left of the name of each piece is given the combination number and card number of the drawing on which the piece is shown. At the right of the name of the piece is given the number of these pieces used on the machine, and the material. Material is signified by initials to save space. Still further to the right is a blank space for checking marks.

The bill of material is first made out by the drafting room in lead pencil on a large, specially ruled form, as shown in Fig. 3 (page 51).

This sheet is taken to the office where a typewritten, exact copy is made of it on a sheet of transparent paper, using carbon paper turned wrong side up underneath. This prints typewriter ink on one side of the sheet and carbon on the other, the object being to make as opaque letters and numbers as possible, for blueprinting purposes.

This typewritten sheet is sent back to drafting room and a Vandyke, or brown, copy of it is made in the same manner as blue printing. This brown print becomes the master copy. From this brown print a number of blue prints are made, the number depending on the popularity of that particular size machine. This print will have a white background and the rulings, numbers and wording will be in blue. The routing clerk can

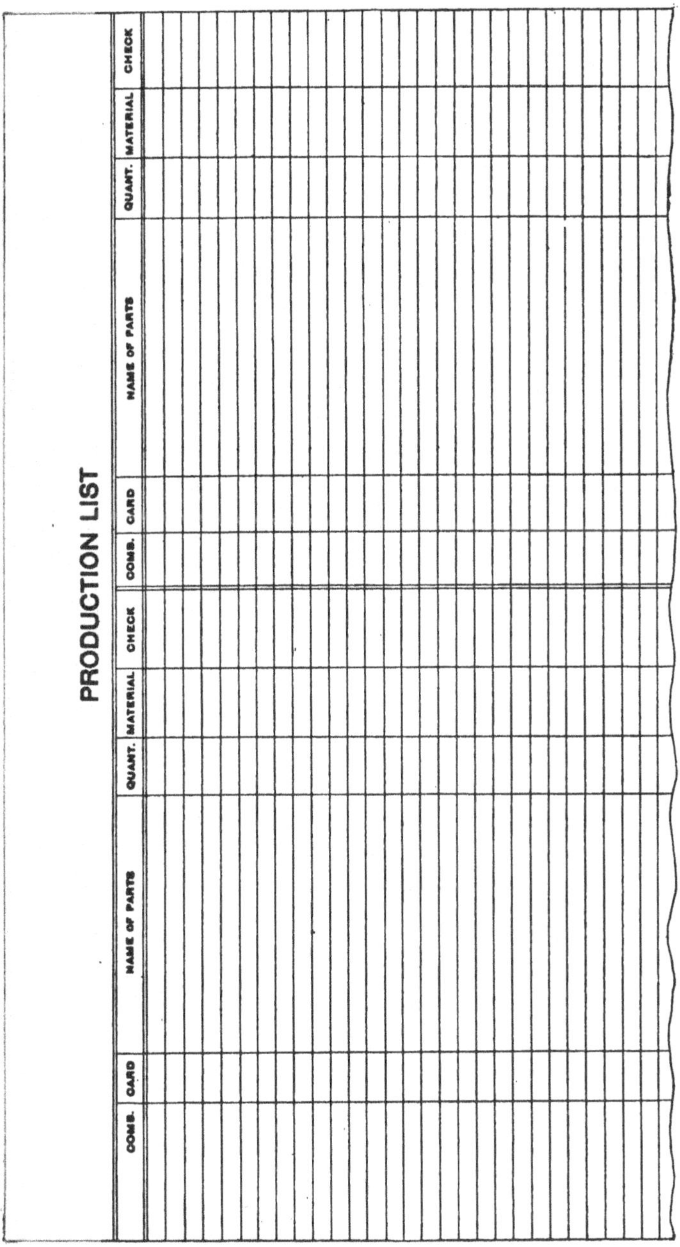
PRODUCTION LIST

COMB.	CARD	NAME OF PARTS	QUANT.	MATERIAL	CHECK	COMB.	CARD	NAME OF PARTS	QUANT.	MATERIAL	CHECK

Fig. 3—Bill of Material. Size, 10⅝x8⅞ In.

make check marks on it on the white background. All bills of material are of a standard size so that they can be filed easily.

When the two duplicate order sheets are received by the drafting room one is filed for reference; the other is pasted on the bottom of the blue print bill of material and sent to the shop routing clerk.

To the shop routing clerk this sheet becomes what the train dispatcher's board is to the train dispatcher. On it he keeps track of all material ordered outside of the factory, all the rough material from one department to another in the factory, all casting orders, and all the machine orders. He can tell by a glance at the bill of material whether a drawing is to be made for a piece, whether material is to be ordered, or has been ordered, whether a pattern is to be made, or a casting is to be made or is in stock in the rough, whether a piece is being worked in the machine shop, or whether it is finished and in stock ready for the erecting. Thus he has complete and accurate control of all parts, and can rush the lagging ones.

Alterations made on the machine while building, all improper workmanship, flaws or any other notes are inserted on the bill of material. These are useful for future reference should trouble arise from these faults.

The test record made out by the testing department is also pasted on this bill of material.

After the machine has been shipped the bill of material goes to the office and becomes the basis on which the cost is figured for the machine. The bill of material is finally filed away as the complete record of the machine, and in after years is used to make out repair part orders for the machine, should any repairs be needed.

The Shop Material Routing Clerk

As soon as the shop routing clerk receives a bill of material for a machine to be built, he sends it to the finished stores department. For each piece they find that they have in stock they stamp their check mark opposite the name of that piece. This is a "Finished Condition" mark.

The bill of material is then sent to the rough stores department. The rough stores clerk stamps opposite the names of those parts that he has in the rough his "rough parts in stock" stamp.

It then is sent back to the shop routing clerk. He looks over it and orders the material for those parts that are not checked, that is, that are still blank. As fast as he writes out the orders for material he makes opposite the name of the piece his own check mark in lead pencil. He uses a different mark for material ordered outside the shop than is used for that ordered in the shop. He looks down the numbers at the left of the names of the pieces, the numbers representing the drawings for the pieces. Any number higher than the last drawing made he checks with a mark that means drafting room.

The object of having the rough stores clerk and finished stores clerk use a stamp for their marks is to identify their O. K.'s from the routing clerk's mark, should any dispute come up later because a certain piece failed to materialize as checked.

By a glance at the bill of material the routing clerk can tell the condition of all the parts. His drafting room check means that a drawing is wanted. A blank space opposite a name means that the piece is not in existence, either in the rough or finished state, and must be ordered. A rough stores clerk check means that the piece is in the rough and must be machined. A finished stores check opposite a piece means it is all complete ready for the erector.

The Casting Order

The casting order is a printed blank along the lines indicated in Fig. 4. The routing clerk completely fills in all spaces except the pattern number and weight, getting all information from the bill of material. The pattern number is put on by the pattern storage man, and the weight by the rough stores man, when he weighs the casting.

Machine to be used on

Size.................... Style ....

Shop No................ Comb. No..............

Make........ castings Material ....

It is shown on drawing No. ....

Card No. ...., and is called the. ....

....

....

Pat. No...................... Wt...............

Fig. 4—Casting Order Blank

The casting order takes a circular course. First, it is sent to the pattern storage man, who gets out the pattern, chalks the number of castings wanted on the pattern, and enters the pattern number on the casting order. The pattern is sent to the foundry, and the casting order goes to the rough stores clerk. This notifies the rough stores clerk to be on the lookout for these castings. As soon as the castings are all made the rough stores clerk weighs them, enters the weight on the casting order, and sends the casting order back to the material routing clerk.

The receiving of the casting order notifies the routing clerk that the casting he needed is now in stock in the casting store room. All this has been done with a minimum of clerical work on the part of the routing clerk, and practically none on the part of the pattern storage man and the casting storage man.

Machine Order

The ordering of a piece to be machined is as follows: A machine order, Fig. 5, is written out in duplicate. The original order and the duplicate are given a consecutive job number for cost-keeping purposes. At the

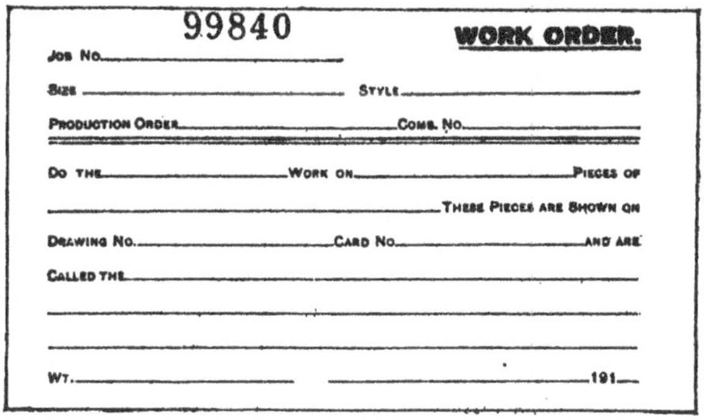
99840
WORK ORDER.
JOB NO.
SIZE ______ STYLE ______
PRODUCTION ORDER ______ COMB. NO. ______
DO THE ______ WORK ON ______ PIECES OF
______ THESE PIECES ARE SHOWN ON
DRAWING NO. ______ CARD NO. ______ AND ARE
CALLED THE ______
WT. ______ ______ 191__

Fig. 5—Machine Order. Original, 3x5¼ In.

same time that the order and its duplicate are written a tag is made out called the job or instruction tag, which is a fac-simile of the machine order. The job number is written on this tag also. All job numbering is done with a numbering machine set to print in triplicate.

The weight of the piece in the rough is placed on the duplicate machine order. The duplicate of the machine order is sent into the cost-keeping department. This is its notification that time will begin to come in to them on this job.

The machine order and tag are sent to the rough stores clerk. This is his notification to deliver the casting or material to the machine shop. He ties the tag on the casting and delivers it to the machines. As soon as it is delivered he hands the machine order back to the routing clerk. This is the routing clerk's notification that the casting is now in the machine department, with the tag tied on it.

When the routing clerk receives the machine order he checks the bill of material, not with a mark as before, but with the job number that the piece is to be machined under. He copies this number from the machine order upon the bill of material. After thus checking, he sends the machine order to the machine foreman. This is the latter's notification that the material has been already delivered to him, properly tagged.

The foreman knows that the tag contains all the information necessary to enable the workman to finish the piece. The size, the style and the order number of the machine of which the piece is to become a part are stated on the tag; moreover, the number of the drawing on which the piece is shown, and the job number to which the workman must charge his time is given on the tag.

After the casting is machined it is sent to the assembling floor with the machine order. The assembler does the necessary handwork, charging his time to the job number. When the handwork is all done, and the piece is ready for the erector, the assembling foreman returns the machine order to the routing clerk, who is thus notified that the piece is finished and on the erecting floor. The routing clerk checks off the name of this piece from the bill of material by drawing a vertical line down through the name. The idea of using the vertical line is that when all the parts for the machine are finished the vertical line will be continuous from top of sheet to the bottom. Any break in it will be noticeable and will call attention to something missing.

The machine order is stamped "finished" by the routing clerk, with the date, and is sent to the cost department. This is its notification that no more time will come in on this piece, and that the cost of the piece can now be computed.

The ordering of the rough storage man to deliver the rough piece to the machines; the notifying of the routing clerk that it has been delivered; the ordering of the machine shop to machine the piece; the giving the workman all necessary information about machining the piece, and also the job number under which he is to work; the notifying the cost department to expect time on the piece; the notifying the assembling department that the piece has arrived there to be assembled; the notifying the routing clerk that the piece is all finished; the checking of the bill of material, and the notifying the cost department that piece is finished is all done with a minimum amount of clerical work on the part of the routing clerk, and none at all on the part of the rough storage man, the machine foreman, or assembling department foreman.

As soon as the bill of material shows that the principal parts are machined and on the erecting floor, and that all the other parts are in a sufficiently advanced state to avoid a possibility of holding the work back, an erecting order is written, which is given a job number and is handled by the cost department in the same manner as if it were a single piece. This erecting job number is written on the bill of material and also on the erector's large descriptive tag that was received from the office at the time the order came in. The erecting order and tag are sent to the erecting foreman. This is his notification to erect the machine, all parts being either finished or nearly finished.

The workmen send their time for erecting to the cost department under the job number on the erecting tag. This tag is tied to the machine at the commencement of the erection and remains until the machine is shipped, and gives complete information to all concerned, even to the testers and shippers.

The testing department makes out a detail test sheet for each machine, which is pasted upon the bill of material by the shop routing clerk. The list of fittings to be shipped with the machine is also pasted on the bill of material.

The bill of material, stamped with the date of shipment, is finally sent to the cost department, and after the cost of the machine, as a whole, is figured, it is filed away to be the complete record of the machine.

In the tool room is filed a master set of bills of material, which is an index to the drawing racks. One detail might be added:

Before beginning erection the erecting foreman notifies the routing clerk to deliver the parts of the machine to the erecting floor. The routing clerk then hands the bill of material to the finished stores man, who forwards the material indicated, the small parts being placed in an open box, chalk-marked with the order number of the machine.

It is a good idea for each person connected with the shop routing of parts to use a system of crude picture marks, dots, dashes and circles to represent the different items on the bill of material; a sort of shorthand that will take up small space and can be quickly written. This will be used for quick memorandum only. Using this system, a large number of notes can be put on a small sheet very rapidly.

On each casting is cast a pattern number and this number is placed on the drawing, which identifies a pattern or casting, and is useful in the pattern storage room, casting storage room, machine shop, erecting room and finished stock room.

For larger plants, the material routing system can be made more elaborate, but in doing so the principle of absolutely keeping clerical work from the foremen must be strictly adhered to.

Foundry System

The following is a system to take care of the work in a foundry:

The bill of material is similar to that used in the system just described, but is more complete. In addition to the number of the drawing that shows the casting, it states the pattern number and the number of core boxes for the casting.

The order for a casting is made out in triplicate. The same information is given as before, with the addition of the pattern number and the number of cores the casting requires.

The triplicate orders are sent to the pattern storage man who gets out the patterns and core boxes. They are sent to the foundry, with the triplicate casting orders.

One order goes to the core maker with the core boxes. This notifies him of the number of cores to make. The other two orders go to the foundry foreman, or his assistant. He gives one, together with the pattern, to the molder, and on the other he enters the molder's number. At the end of the day he gives the order to the rough storage or casting storage man. This is the notification to the casting storage man to watch for these castings.

If the molder fails to make the full number of castings called for on an order, or if there is a shortage on account of bad work, the casting storage man notifies the foundry foreman in writing, giving the molder's number, and the name, etc., of the casting. This is a copy of the casting order, but is on different colored paper. Constant shortage from one molder shows strongly. These slips are kept as records of the molder's bad work.

In the machine shop when work falls behind the machine order slip is called in and replaced by a similar slip of bright red. On this slip is given the date on which the piece must be finished. The same change is made with the job tag—a red one is substituted for the white. The red tag must be used sparingly, otherwise it will lose its effect.

Taking Care of Patterns

If a plant is large and has many patterns to take care of, a good system is to number the tiers of pattern storage shelves and letter the shelves in a tier. Make out a book of pattern numbers. Enter opposite each number the shelf number and letter. The patterns are stamped with their shelf number and letters. This system makes a firm completely independent of its pattern storage man. Any one can run the pattern storage department without previous knowledge of the patterns or their location in the storage rooms.

Short Cuts

Number the requisition orders for material that the plant buys from outside firms thus:

Start all January orders with a 1; February with a 2; March with a 3, etc. The first order in January will be 1-1; the second 1-2; the ninth will be 1-9; the tenth will be 1-10; the ninety-ninth will be 1-99; the hundredth will be 1-100. In actual practice omit the dash. The above numbers will then be: No. 11 for the first; No. 12 for the second; No. 19 for the ninth; No. 110 for the tenth; No. 199 for the ninety-ninth, and No. 1100 for the hundredth.

The advantage of this system is that the order number tells the date as well as the number.

An ingenious system of tabulating in a small space numbers that run in the thousands, such as the shop numbers of machines under construction, is to use a cross-ruled sheet, one page for each one hundred numbers. The columns across the top are headed 0, 10, 20, 30, 40, 50, 60, 70, 80 and 90. Down the left-hand edge they are numbered 0, 1, 2, 3, 4, 5, 6, 7, 8 and 9.

Number 22678 would be in the square where number 70 vertical column crossed number 8 horizontal column on the page of numbers running from 22600 to 22699, thus:

Sheet of all numbers from 22600 to 22699

	0	10	20	30	40	50	60	70	80	90
0								..		
1										
2										
3										
4										
5										
6										
7								..		
8	......	......	......	......	......	......	......	x		
9								..		

The next sheet above this would be 22700 to 22799.

To record all the numbers from 22600 to 22699 in full, in the regular way, would take over three times the space, and about five times the time, with a greater chance of error than with this system.

The making of a record running into the thousands with this system is the mere jotting down of a cross or other check mark, in the correct square. These squares can be made large so that more than one kind of a record can be made in each by using different kinds of check marks.

ARTICLE VI

ACCURATE DELIVERY PROMISES

How to Ascertain the Time to Manufacture an Article Involving Foundry and Machine Shop Operations—Use Made of Complaints

THE superintendent can give an accurate delivery promise by making a time schedule of that casting on a given job which takes the longest time to get through the plant. Other parts of the job need not be considered, if it is assumed that they will arrive at the assembling floor before this controlling casting gets there, and steps are taken to insure that they will. The time required by this casting is the shortest possible time in which the complete job can be turned out.

In busy times the superintendent should add to this time an amount representing the time that the casting under consideration has to wait to get into the various machines. To do this he must know how far behind the plant is in its work. If the total possible tonnage output of the plant per month is known and compared with the total tonnage of the orders on hand to be filled, the number of months, or the fraction of a month that the plant is behind is shown. This time, added to the shortest schedule time, will give the total time any given order will require in the shop.

The material routing clerk sees that all the other parts are routed through the shop so that they arrive at the erecting floor ahead of the controlling piece. At the time the controlling piece arrives on the erecting floor an erecting order is written by the routing clerk.

He passes the bill of material to the finished stores department, which checks all the material to see that no part is missing. If a part is missing the material routing clerk is notified and he sees that the part is rushed to the erecting department so as not to delay the erecting.

Determining the Accumulation of Work in Departments

It is a good plan to mark the weight on each order. File the order sheet under the name of the department in the plant in which the commanding casting is. Thus one portion of the order sheets will be under the office, drafting room and pattern shop division of the file, another portion will be under the foundry, another under the machine shop and still another under the erecting division of the order file. It is well to divide the machine shop portion of the file into two divisions one representing the first machines that the commanding casting goes to, and the other representing the later machines.

Once a week the weights of all the orders in each division of the file are added up and a table is made giving the weight of orders in each department. Then by dividing the pounds of orders in each department by the average daily tonnage capacity of the plant, the number of days' work ahead in each department is shown. This indicates whether or not more men should be put to work in any department. Keep the correct number of men in each department. If a plant has more men than work, the output per man will drop and the cost rise. If the work accumulates, the customer will suffer.

When the plant gets far behind in its orders it is a good idea at intervals—say once in two weeks—to make a list of all orders on hand, arranging them in the order of their shipping dates. Place the most urgent order at the top of the list, the next second, etc. Give copies of this list to each of the foremen and straw bosses.

Such a system will insure shipments being made on the promised date, assuming, of course, that judgment is used in making up these dates. Of course, this list will be continually rearranged. Some orders will be inserted ahead of others. Where a special case comes up, even a new order may be inserted near the head of the list.

Figuring the Time of Completing an Article

The system may be made clearer by an example. Suppose, by observing the weights of pumps shipped during a stretch of time when the plant is pushed to the limit, the output is found to be 10,000 pounds per day. The weight of each pump ordered is placed on the order sheet. If the weight is unknown its estimated weight is placed on the sheet. These order sheets for pumps still to be built are filed, as mentioned elsewhere, in a file which is divided into the following department sections: (1) office, drafting room and pattern shop; (2) foundry; (3) boring lathes; (4) other machines than boring lathes, and (5) erecting.

On the first and fifteenth of each month the weights on these order sheets are added up for each department. Suppose these weights are as follows: 48,000 pounds of pumps held up by office, drafting room and pattern shop; 25,000 pounds of pumps in the foundry; 100,000 pounds of pumps waiting to have the cylinders bored; 160,000 pounds of pumps the cylinders of which have been bored, but all of the controlling parts of which are not as yet on the erecting floor; 160,000 pounds of pumps in the erecting department· With an output capacity of 10,000 pounds per day, the plant then has 4 4/5 days' work held up by office, drafting room and pattern shop, 2½ days' work in the foundry, 10 days' work by the boring lathes, 16 days' work in other machines than boring lathes and 16 days' work for the erecting floor. From these figures it is a simple matter to tell how long a pump will be delayed in the plant by the orders that are ahead of it, if it takes its turn.

Providing a Date Schedule for Rush Orders

The time it will take to rush a single order through the plant ahead of everything, regardless of the other orders, is ascertained as follows: Each pump has some one part on it that takes longer to get through the plant than any of the other parts. This is the part that controls the shipping date. Suppose it to be a pump cylinder. A schedule is made out on this part thus:

January 2—Cores made.
January 3—Casting poured.
January 4—Casting cleaned.
January 5—Sunday.
January 6—Bore.
January 7—Mill.
January 8—Drill and tap for seats.
January 9—Clean and seat.
January 10—Erect.
January 11—Erect and test.
January 12—Sunday.
January 13—Ship.

A copy of this schedule is given to each of the foremen and one to the material routing clerk. This estimate of the time allowance for each step in the work in the schedule is obtained from each of the foremen, so that the schedule is the foreman's own estimate of the time it takes to get the controlling piece through his department. This being so, the foreman sees to it that the piece goes through on time.

Utilizing Complaints from Customers

Whenever a customer makes a complaint about a flaw or bad workmanship, give a copy of the letter to each of the foremen, testers, inspectors, etc., that are implicated. It will have a stimulating effect, and will put these men right in the game. An improved product and a reduction in the number of kicks will result.

ARTICLE VII

COST KEEPING IN A FACTORY

A System Designed to Give Accurate Information with Special Effort to Minimize the Clerical Labor Required of Shop Employees

EACH lot of pieces to be made is given a job number. The workman gets the number from the job tag or instruction tag which is tied to one of the pieces of the lot. The foreman can get it from his machine order, and the cost department gets it from a carbon duplicate of the foreman's machine order.

Each workman is provided with a pad of blank time tickets with the hour and quarter-hours of the day printed in a straight line across the bottom. He places a cross on the hour or quarter-hour mark on the ticket that represents the time at which he starts the job, and another cross on the hour or quarter-hour mark showing the time at which he finishes the job. He draws a connecting line between these two crosses. He also enters on the slip the job number, the total number of pieces in the lot (quantity) and his own clock number and the date. If he is a machine man, he also enters the number of his machine.

He further checks off on time slip the operation that he performed, whether it be lathe, planing, milling or drilling work, if he is a machine hand, or tapping, scraping or studding if he is an assembler. This checking consists of drawing a line through the word. The system is simple to explain to the workman and simple for him to follow, for there is no writing connected with it.

The claim is sometimes made that the time would be more accurately taken by a clock, which automatically stamps the time of starting and finishing a job. The men should not leave their machines every time they start a new job. It must be remembered that cost-keeping is of secondary importance. The profits are made at the points of the tools, and it is of the utmost importance to keep the men at their machines. That output is the main object must never be lost sight of. A firm with an enormous output per man and no cost system is better off than one with a fine cost system and a low output per man.

For each new job, the workman makes out a new time slip. At the end of the day the quitting time is marked even if the job is not done, and all the time slips for the day are sent to the cost department. The next morning the workman makes out a new time slip for the unfinished job, which will, of course, have the same job number as that of the day before. If the workman's starting time on one job does not exactly correspond to the finishing time of the previous job, the cost department will notify him, and the proper correction will be made.

A different colored time slip is used for each department to enable the cost department to separate at sight the slips of the various departments from each other. On the time slips there is also printed, for special work, a list of those departments that are likely to have work done for them, such as office, machine shop, foundry, pattern shop, yard, power, etc. To reduce the size of the time slips the department names are abbreviated, and a number of instruction sheets are posted under glass throughout the plant, explaining these abbrevia-

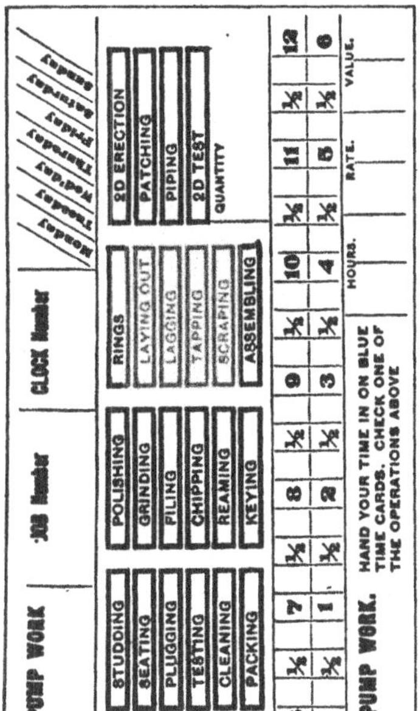

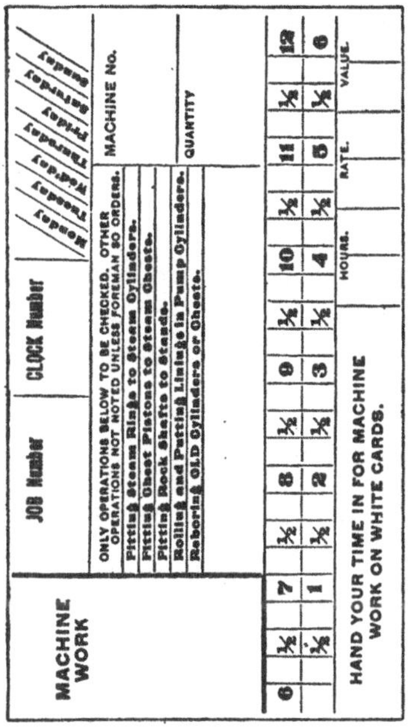

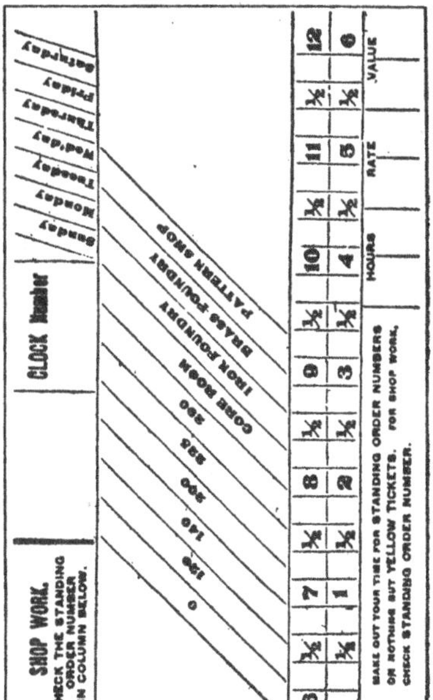

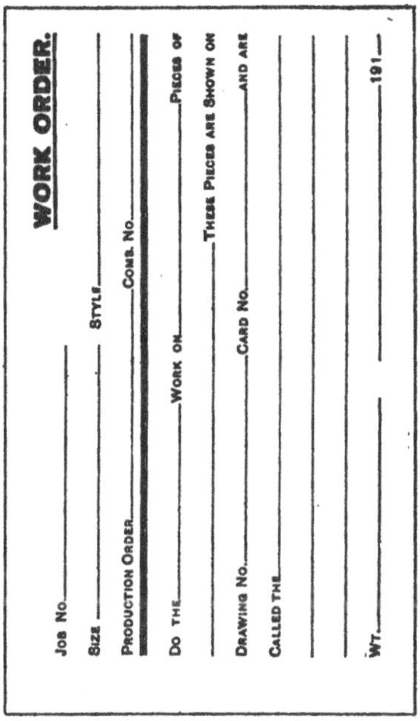

Fig. 6—Time Slip for Assembling Work—Productive Labor
Fig. 8—Time Slip for Non-productive Labor

Fig. 7—Time Slip for Machine Work—Productive Labor
Fig. 9—Work Order Form for Stock, Special Parts or Repairs

tions. When a man does a repair job for one of these departments, instead of writing a job number, he draws a line through the name of the department for which the work is done. The cost department then charges his time against that department.

In cost-keeping there must be a checking or proving that will detect any mistakes made either by the workman or in the figuring of the costs. The following is the method of checking used by the shop with which the author is connected.

Checking up Time Slips

1. All the time slips are assorted according to clock numbers (employees' numbers).

2. To detect the failure of any workman to send in his time slips, the clock numbers on the time slips are compared with a previously prepared list of clock numbers representing the men who should send in their time on slips. As soon as a new man sends in time slips, his name is added to the list.

3. Each day a list of those employees who should have sent in time slips but who failed to do so is made out. From the list those who were absent are struck off.

4. This list is then sent into the factory, and the missing time slips are collected.

Checking Hours and Wages

1. The hours and quarter-hours, as put down by the workman in the shape of two crosses connected by a line, are entered in a space for this purpose on the time slip. This entry represents the time that was taken to do the job. For simplicity, no time is figured for less than a quarter of an hour. The amount of wages for the time represented by the slip is entered on the slip.

2. The total of each man's wages for the day, as shown by the time slips is compared with his wages for the day, as placed in the time book by the time-keeper. Two girls work together on this checking, one handling the slips and the other the time book.

If the amounts correspond it shows that no mistake has been made by the workman, the cost-keeping department, or the wage pay department. Thus: the workman has made out his time slips correctly; his jobs do not overlap or fall short of each other; the starting time on his new job is the same time as the finishing time on the previous one; the cost-keeping girl has computed the graphically marked time of the workman correctly, and set it correctly down on the slips; the cost-keeping girl has calculated correctly the amount of pay due for the computed time and correctly entered it on the time slips. The time-keeper has correctly figured the man's time and wages and entered them correctly in the time book.

If the amounts do not correspond the mistake is located and corrected. If it is the workmen's mistake the shop is notified, and the error is rectified.

Separating Department Charges

The time slips for each day are separated according to the following departments: Machine, assembling, erecting and non-productive. The non-productive slips are further separated into departments. The wages and hours are totaled in each of the classes and sub-classes of time slips. Specimen slips are shown in Figs. 6 to 9.

A daily sheet is then made out on a printed form, Fig. 10. The above totals are all entered on this sheet, which gives the total number of hours for each department, and also the total wages for each department. It further gives the total amount of money spent for material for each department, and all other expenses connected with that department. That is, any money spent during the day, no matter for what purpose, whether for pay-roll, material, insurance, taxes or work done by an outside firm, is placed on this sheet, charged to some department or to some division of the business. The items in these daily sheets added together make the monthly sheets. From them is made out the yearly sheet.

The principal object of this sheet is to show the proportion which the overhead, or non-productive, expense bears to the productive expense. This is the "pro-rate." Besides the pro-rate, this sheet gives the cost in each department, and thereby is useful to the manager or superintendent in keeping close watch of department costs.

At intervals the amount of non-productive labor sent to the office will have to be looked into, as there is always a temptation to send in non-productive time which could be charged to jobs. The writer knows of one case where the non-productive labor rose to 62 cents per hour, and was decreased to 42 cents per hour by carefully scrutinizing all non-productive time slips.

After all the time slips have been separated according to departments, and non-productive hours and money have been entered on the daily sheet, the total number of hours are added together and compared with the total number of hours in the time book. These must correspond. This is the check to show that no person on the pay-roll has been missed, and that no labor cost that should have been charged to a department has been overlooked.

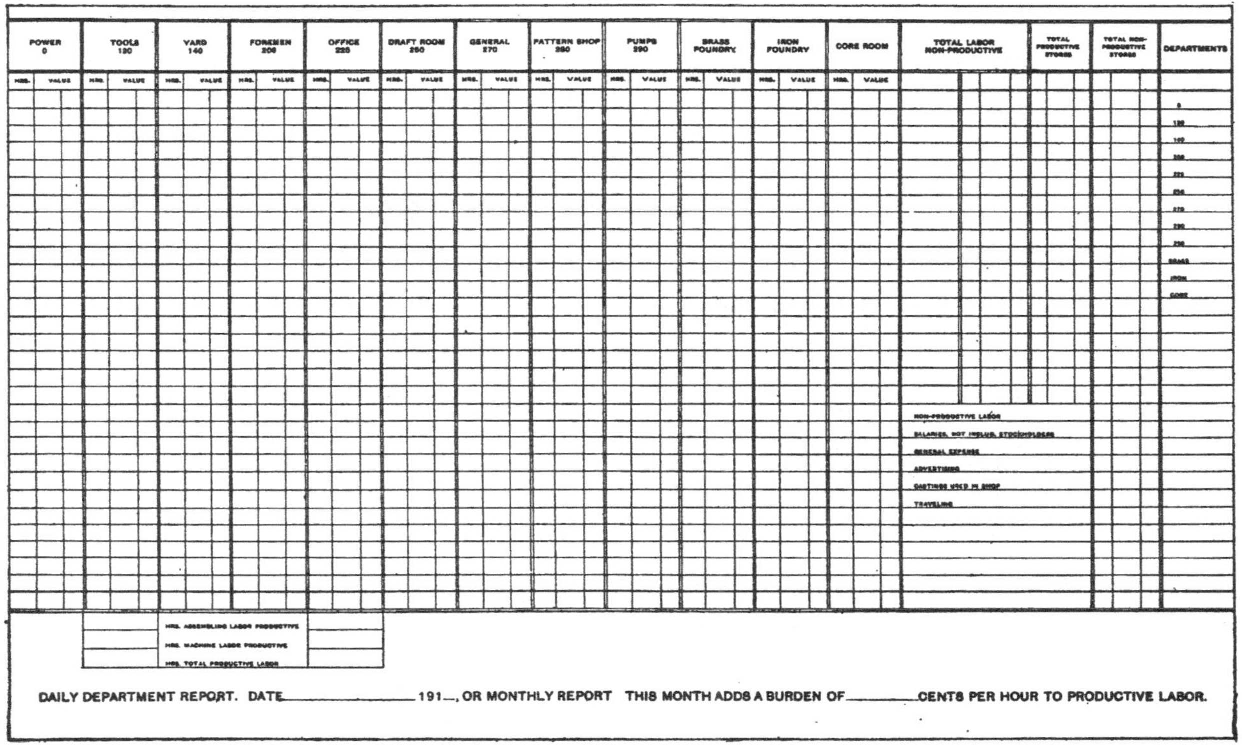

Fig. 10—Reproduction of a $10\frac{3}{4}$x17 In. Sheet Used for Distributing Daily, Weekly and Monthly Non-Productive Labor Charges

Apportioning Time to Job Numbers

All the time slips are filed according to job numbers. The girl while doing this sees that the number of pieces worked on, as marked on the time slip, is the same as the duplicate machine order that was received on the job when it was started. When there is a difference, the time slips are sent to the shop for correction. This acts as a check. Sometimes workmen put down the wrong job number and the discrepancy in quantity will disclose the mistake.

These time slips remain filed until the notification comes in that the job is finished.

Ascertaining the Cost of Individual Parts

The cost of an individual part of a machine is called the flat cost of the piece, and the cost card is known as the flat cost card. Each one of these cards has space on it for entering the itemized time and cost of the piece four times, which gives a chance for comparison. A new flat cost card for a certain piece is never made out until the last one is completely filled.

These flat cost cards are filed in boxes in the order of their combination numbers. The name of the piece, the combination number, and the drawing card number are written at the left-hand edge so that the cards can be rapidly handled. These items are copied from the machine order duplicate that has been filed with the shop-workman's time slips.

The combination number is the design number of one size of a pump as a whole, as it was built, and differentiates the size, style and design of the pump down to the smallest detail from all other designs, sizes and styles. Combination No. 1900, for instance, by referring to the master production list or bill of material, shows exactly what parts were used on the pump. No pump with combination No. 1900 will be built any differently. If a slight change in the shape of any of the parts should be made on this combination a new combination number would be given. The material may be changed in certain parts without a change in the combination number. Rubber valves may be changed for brass or a brass piston rod may be changed to steel. Or even a brass cylinder may be placed in the pump instead of cast iron, provided exactly the same design and pattern is used. Changing the material, however, without changing the combination number is allowed only on a few certain parts that are always specified by the customer in his original order.

When a piece, or a lot of one kind of pieces, is completely finished in the shop, the shop sends in the original machine order to the cost department stamped "Finished." When the cost clerk receives this original machine order stamped "Finished" he takes all the shop-workman's time slips, which have the same job number as the machine order, to his desk. These shop-workman's time slips have been gathered up daily from the workmen and have been accumulating in the cost department files while the job was in progress.

The clerk then takes to his desk the flat cost card, Fig. 11, having the same combination number, card number and name of piece on it as the machine order which was marked "Finished." If no such card be found in the files, or if the card be full, a new card is started. The information on the shop-workman's time slips and the machine order are now entered on this flat cost card. The date, the quantity worked on and the job number are placed at the top of the card. The material of which the piece is made and the cost rate at which the material is figured are placed next. This last is done so that if material rates change in the future, the cards can still be used by making the correction. Next is entered the weight of the piece. All these items are copied from the machine order.

The weight entered is the rough weight before any work has been done on the piece. No credit is given for stock removed. When the piece is made from bar stock, the dimensions of the stock are entered on the machine order instead of the weight. The cost department has tables for transposing these dimensions to weights.

The sequence of operations then is:

1. Separate from the others all the shop-workman's time slips that have notes on them referring to bad workmanship or flaws in material. A special note is made of this extra time and extra expense, not on the flat cost card, but on the final cost of the machine as a whole, because it is assumed that this particular extra expense will not happen again. What is wanted on the flat cost is the true cost of the piece as it would have been had the work gone smoothly. This extra time and expense must not be thrown away, and therefore it is charged to the finished machine as a whole.

2. Separate from the others all the shop-workman's time slips that have special notes on them referring to the number of pieces thrown out as bad. These notes are put on the flat cost card of the piece in red ink.

3. Separate the shop-workman's machine time slips from the erecting and assembling slips. They are of different colors.

4. Separate the machine time slips so that all the slips of one man are together.

5. On the flat cost card, in the proper column, lathe, planer, shaper, etc., enter the number of the man's ma-

CONS. CARD.	MONTH YR.	MONTH YR.	MONTH YR.	MONTH YR.
No. of Pieces	Start Fin.	Start Fin.	Start Fin.	Start Fin.
Job Number				
Material at Rate				
Total Weight & Rough Value	Lbs.	Lbs.	Lbs.	Lbs.
	Mchn. Hrs.	Mchn. Hrs.	Mchn. Hrs.	Mchn. Hrs.
Eng'n Lathe				
J. & L. "				
Chuck "				
Brass "				
Polish "				
Boring "				
Planer, Shaper				
Milling				
Drill				
Tapper				
Bolt Cutter				
Machine Total	Hrs.	Hrs.	Hrs.	Hrs.
As'b'l Total	Hrs.	Hrs.	Hrs.	Hrs.
Flat Cost Total	Hrs.	Hrs.	Hrs.	Hrs.
One Piece: Machine	Hrs.	Hrs.	Hrs.	Hrs.
One Piece: As'b'l	Hrs.	Hrs.	Hrs.	Hrs.
One Piece: Rough Stores	Lbs.	Lbs.	Lbs.	Lbs.
One Piece: FLAT COST	Hrs.	Hrs.	Hrs	Hrs

STYLE — SIZE — PART

Fig. 11—Example of Flat Cost Card, 4 x 6 In. in Size

chine, the hours he spent on the lot of pieces, and also the wage cost on the lot.

6. Assort the handwork, or assembling slips according to the different hand operations, such as laying out, filing, tapping, studding, polishing, chipping, scraping, cleaning, piping, keying, testing, etc. Enter the number of hours spent and the wage cost for each of the above hand operations in the proper place on the flat cost card.

7. Total the machine hours and wages on the flat cost card and enter the amounts in their proper places at the bottom. Do the same with the assembling time. Add the total machine time to the total assembling time, and divide by the number of pieces in the lot. This is entered in the place marked "Hours on each piece." The same operation is performed with the wage costs. Finally this cost card is filed in the flat cost card file according to its combination number and card number.

All erecting time on a machine as a whole is charged to the erecting job number of the machine. An erecting cost card is made out in the same way as the flat cost card for a piece.

The flat cost per pound should be placed on each flat cost card. Long time taken on a piece, wrong time sent in on a piece, mistakes in figuring the costs of labor and material then will be easily discovered. The design or the method of manufacture will be changed on all pieces that show a cost over a certain rate per pound. A marked reduction in the cost of the product can be made if the above is carried out.

Ascertaining the Cost of a Complete Machine

When the machine is finished, the bill of material is sent to the cost department marked "Finished." The cost department gets from the drafting room an exact duplicate sheet of this bill of material and the following routine ensues:

1. The cost department pastes at the right-hand edge of the blank bill of material a large ruled and printed form (Fig. 12) made especially for the purpose of cost figuring.

2. All notes that the shop has made on the original bill of material are transferred to the blank bill of material. Next, all the machine job numbers are transferred from the shop bill of material to the blank list. Then all the items mentioned as being sent on the fitting list are entered.

3. All the flat cost cards are taken out of the files pertaining to this bill of material. Only those are used that have the same job number as the job number given on the bill of material. Where no job number is given it shows that the shop got the piece from finished stock. In this case the latest cost on the piece is used.

Next, all weight items from the flat cost cards for each piece and also hours' labor for each piece are entered. The columns are then added and the totals are entered. The adding machine is used on the long additions to save expense and to insure accuracy.

The final condensed cost of the completely finished machine is put on a card about 4 x 6 inches. On it are the following items:

Size and style of machine and order number.
Machine labor value.
Assembling labor on parts, value.
Erecting labor.
Machine hours.
Assembling hours.
Erecting hours.
Total productive labor hours.

The total productive labor hours are multiplied by the pro rate per hour, which gives the overhead expense that the machine has to carry. This value is entered on the card. The following items also appear:

Value of purchased material.
Value of rough material.
Weight of cast-iron castings.
Rate figured at ....cents.
Weight of brass castings.
Rate figured at .... cents.

These items are put on so that if the rate changes the corrected cost can be figured.

Weight of other material.
Rate figured at .... cents.
Painting and skidding.
Freight.
Weight of finished machine.

The hours of erecting can be sub-divided if the firm builds a machine that sometimes has extras, that at other times are omitted. The erecting of these extras can be kept separate and entered as separate items on the card. On the back of the card are written details that tell how the machine is built.

These main items are copied direct from the totals at the bottom of the large bill of material cost sheet, Fig. 12, excepting the erecting time and value. This is taken from the erecting flat cost card. The pro rate is figured from the yearly cost sheet, as previously mentioned.

All special notes are added to the cost card, such as time and value of extra machine work that had to be done after the parts were considered finished; testing and cleaning, second erection, second testing, pattern time and pattern labor cost. Where a machine is made by changing the pattern of some other machine, the cost of alteration should be added to the cost of the machine, as the customer should pay the expense of pattern changes.

COST SHEET
OF
ONE COMPLETE APPARATUS

Size 5½ x 7 x 7 Style D F W A Comb. No. 165 Order No. 18,255			PURCHASED FINISHED MATERIAL	ROUGH MATERIAL	VALUE MACHINE LABOR	VALUE ASSEMBLING LABOR	HOURS MACHINE LABOR	HOURS ASSEMBLING LABOR	WEIGHT CAST IRON	WEIGHT CAST BRASS	WEIGHT STEEL	WEIGHT PUR. MAT.
165	1	Steam Cyl.										
165	2	Steam Piston Head										
4	1	Steam Cyl. Head (Fit to Cyl.)										
165	2	Steam Piston Follower										
165	2	Inside Packing Rings										
165	2	Outside Packing Rings										
230	1	Steam Chest Cover										
165	1	Slide Valve										
230	6	Valve Rod										
230	5	Valve Rod Head										
6	4	Valve Rod Stuff Box										
6	4	Stuff Box Gland										
6	4	Stuff Box Nut										
230	6	Piston Rod										
3	3	Steam Piston Rod Stuff Box										
3	3	Stuff Box Gland										
3	3	Stuff Box Nut										
165	3	Crank Shaft										
230	6	Crank Shaft Cap										
165	5	Links										
230	6	Link Stud										
230	6	Link Bolt										
230	6	Link Slide										
230	6	Link Slide Plate										
230	6	Crank Brass										
230	6	Crank Brass Slide										
165	5	Eccentric										
165	5	Eccentric Arm										
165	5	Eccentric Strap										
165	5	Rocker Arm										
165	5	Rocker Arm Upper Pin										
165	5	Rocker Arm Lower Pin										

Fig. 12—Part of Blank Used for Ascertaining the Cost of a Complete Machine

Painting and skidding are taken from a separate flat cost card made out in the same manner as the other flat cost cards, and entered on the condensed cost card. The total cost is then summed up and entered on the same card. The selling price also is put on the cost card as is the profit and also the percentage of profit.

Use Made of Cost per Pound Records

After the total cost has been summed up the cost per pound is entered on the card. This is based on the net weight; that is, the weight without the skids. The cost per pound is useful as a rough check on costs. If the cost per pound on any particular machine is higher than the cost per pound of other sizes of the same style of machine, it is at once evident that something is radically wrong in the manufacturing, which must be corrected. The cost per pound is useful also in making estimates for new work. The setting of the selling price on a machine, the cost of which is not known, is a delicate undertaking. The price must not be so high that the order is lost, nor so low that money is lost.

Accurate selling prices can be made on machines where the cost per pound is known on similar machines. Where costs are hazy there is a temptation to sell at or below cost. A known cost of manufacture stops this.

The cost system described above is simple. Two or three girls can take care of the time-keeping and the cost-keeping for a firm employing two hundred men, and turning out a diversified product. No part of the cost-keeping system is carried on by any one in the shop. The cost-keeping is simply tacked on the shop routing system.

Figuring Pro Rate or Overhead Expense

From the yearly expense sheet, made out by the cost-keeping department, the pro rate is made.

One of two systems can be used for this charge. Either take a certain fixed percentage of the productive cost (labor plus material), make this the pro rate and add it to the productive cost, and call this the final cost, or add a certain fixed amount, so many cents, for each productive hour worked on the product. The amount to be added for each productive hour is taken from the previous year's proportion of total overhead expenses to the total number of productive hours for the whole year. For instance, suppose that total overhead expense for the entire year divided by the total number of productive hours was 45 cents. Then to get a true cost, to the flat cost of each productive hour 45 cents must be added to carry all the other expenses. The productive hours are those hours actually charged by the workmen to a particular piece or particular machine. The overhead expense means every charge of any kind, no matter what it be, above the productive labor and the material that is directly charged to the product.

This last system is the only correct one to use. It favors those jobs where the hours of labor are few as compared to material cost. It strongly favors those jobs that are completed in very short times. It thus encourages rapid work and makes it impossible to employ slow men in a plant.

A plant using the first mentioned pro rate system can increase its capacity by adopting the second system, because the second system favors work that takes few productive hours as compared to the material cost and discourages the taking of orders having labor cost that is high as compared with the material cost.

A plant run to its maximum capacity has only a fixed number of hours that its men can put in on the work. Any system favoring the taking of orders requiring but an hour's work will increase the output of the plant and likewise the profits. A firm should make its main profit on the labor of its men, machine tools and plant and not on the material it uses. The labor item is the output restrictor of a plant.

Firms engaged in a class of work where a large amount of material is finished by a small amount of labor can make money when selling their output at a profit of 5 per cent., because their tonnage output per year is enormous, whereas, a firm whose main expense is labor would go into bankruptcy at the same low percentage of profit, their output being limited. The percentage of pro rate or overhead expense in a plant will increase as the output per workman increases, because:

1. The keeping of tools in a high state of efficiency and the employing of efficient foremen must increase expenses. Moreover, hard driven tools require constant repairs and replacement, and, furthermore, labor-saving devices, such as jigs, fixtures and special small tools, are expensive.

2. All reduction in productive labor means a reduction in the productive labor force. Each producer must carry more of the overhead expense as the force is reduced. It is possible to take a plant whose output of finished machines averages 44 pounds per day for each productive workman, and increase this average to 110 pounds per man for each day's work by improvement in design, changes in the amount of finish and methods of doing the work, by employing good workmen and requiring them to turn out a large output.

The overhead expense will be less on each machine turned out than it was before the plant output was raised, but compared to the productive labor expense, which has been reduced in amount, the pro rate will show up proportionally higher, all on account of good management. Take the imaginary case of a firm equipped and managed so well that its whole output was turned out by one productive workman. This one man would have to carry the whole overhead expense. It stands to reason that his pro rate per hour would be something enormous.

The best way to watch the overhead expense, in order to keep it down, is to compare the year's total expenditure for each item with the corresponding expense of the preceding year. When considering whether it would be cheaper to buy certain parts of the product already finished or make these parts in the shop, use the flat cost, without the pro rate added. The pro rate is a charge that the customer must pay so that the overhead expense of a plant may be taken care of.

ARTICLE VIII

PATTERNS WHICH SAVE MACHINING

Things to Be Considered in Pattern-Making to Minimize Investment in Patterns and an Unwise Amount of Work in the Machine Shop

THE design of a plant's product often determines whether or not the plant will make money. A design which involves complicated core work in the foundry, difficult and close machine work, or which is inconvenient to assemble will eat up all the profits by a high labor cost. The proportion that the labor cost bears to the material cost on a machine must be known in order to design intelligently. This makes it possible to judge whether or not an increase in material that decreases labor will pay.

When Not to Save Metal in the Product

For instance, thicken a casting if this saves enough time and lost castings in the foundry to more than pay for the extra metal. On the other hand if the value of the material saved will be more than the expense of the extra labor incurred, it may pay to design so that hand-work is done which would have been avoided had more metal been allowed. This would be true of brasswork where material cost is a big item. In considering costs in this light, the overhead expense must not be added, as that is merely a charge that the customer must pay to take care of the running expenses of the plant.

Wherever possible, machinery should be designed on the unit plan. Divide portions of the machine into units. Use one unit on as many different sizes of machines as possible. These units can be manufactured in lots and producing methods which will result in economy can be worked out in great detail.

The designer must consider not only the strength of the machine he is building but must keep his eye on the pattern shop, the foundry and the machine shop all the time. He must so design that the work done in each will decrease to the greatest possible extent the labor cost in the next following department. The patterns should be made to save the molder's time. Likewise, the casting should be such that no more machining need be done than is absolutely necessary.

Pattern Drafts

Allow 4 degrees draft on the outside of all patterns. Show this on the drawings. Do not depend on the pattern shop to get it. Allow 10 degrees draft on the sides of a cast hole made by green sand. See that these holes are brass bushed or brass lined in the pattern. The sand will draw smoothly and the cast hole will need no drilling or filing to clean out bumps or fins. Unless these rules are followed it will cost more to cast holes than to drill them.

The words "Give abnormal draft to all patterns" should be painted on large signs all over the walls of the drafting room and the pattern shop. They should be drilled into every one who does any pattern or drawing work. If the patterns are designed and built with big draft, there will be a saving made in the foundry that will pay the draftsmen's salaries many times over. Besides reducing the cost of molding, good draft will insure the casting being absolutely true to the pattern, because all the handwork, patching of the mold, etc., will be eliminated. This will permit a great reduction in the amount of finish to be allowed on castings, which often means a reduction of one-third in the machining time. A foundryman once said: "I want so much draft on a pattern that I have to lay pigs of iron on it to keep it from jumping out of the sand of its own accord."

Designing with an Eye to the Cores

Design so that most of a pattern is in the drag and the least amount in the core. The core part of a casting costs 3 cents per pound; the drag part only 2 cents. Get as much of the casting at 2-cent rate, and as little at the 3-cent rate as is practicable. Never design a piece with underarm (parts of pattern that have to be drawn out of the mold horizontally). Such a casting will cost 4 or 5 cents per pound. A green sand core that the pattern leaves must be in the drag part of mold.

Never design so that cores will have to be hung on the sides of a mold. It takes as long to secure these cores as it does to make an entire mold. Design such cores as port cores, of as nearly a square section as possible. Avoid the extreme flattened section. The square section cores are less trouble in the foundry because they are easily rodded and vented, and they can be handled with less danger of breaking.

In many places a cast bolt slot will replace a cast bolt hole. If slots are used, the parting of the mold should come at the end of the slot, not half way down in the slot, else a shift of the mold would narrow the slot and give the assembler handwork.

Cored holes that afterward have to be drilled, unless they are quite large or deep, are bad. With the modern high speed steel drills, the relief of the hole by coring is of little help. The sand that is burned into the casting may necessitate regrinding the drill every five holes.

The grinding time on the drills will more than equal the time saved by the cored clearance holes, and omitting the coring will reduce the molding cost.

One Pattern for a Number of Shapes

Where a single pattern is used to make a number of different shapes of castings, have that part of the pattern that receives the change made separate from the body of the pattern. It should be doweled on and also held with screws. A pattern of this character would be for work that does not come up often enough to pay for a separate pattern for each class of casting. The core boxes may be separated in the same way at the same point. With this arrangement only a few minutes work is needed in order to change from one pattern to another.

The usual way of making such changes is to hunt through a mass of pieces that were used before, and nail them to the pattern, thus gradually ruining it. The mass is sent down to the foundry with a bunch of stop-off pieces, to set the molder and the foundry foreman worrying over the hieroglyphics of chalk marks and stop-offs. The result is a bumpy casting, costing 4 or 5 cents per pound, instead of a fine smooth casting, costing 2 or 2¼ cents per pound, and the chances are that the casting will be lost from dirt, drop out or wash.

Allowances for Finish in Machining

The designer can save the machine shop lots of work by dimensioning the patterns so that there is no great excess of metal to be machined off. The following allowances for finish are about right:

Cylindrical work where the surface is crossed by the molding parting.—A round bar, cast horizontally in a mold, is a good illustration. In this style of a casting there is a chance for mold shift.

3/32-in. finish on a side of pieces 1½ in. in diameter.
3/32-in. finish on each side of pieces up to 3 in. in diameter.
1/8-in. finish on a side of pieces 6 in. in diameter.
1/8-in. finish on each side of pieces up to 8 in. in diameter.

If the piece is long, and, for this reason, likely to be out of true in the rough, double the above allowances for finish.

Surface not crossed by a parting, such as a piece cast on end.—A shift on this casting is impossible.

3/64-in. finish on each side of piece 1½ in. in diameter.
3/64-in. finish on each side of piece 3 in. in diameter.
1/16-in. finish on each side of piece 6 in. in diameter.
1/8-in. finish on each side of piece 12 in. in diameter.
3/16-in. finish on each side of piece 40 in. in diameter.

On small and medium-sized cylindrical castings that are cast on end the pattern has to be given draft. It is best to leave no finish at the small end of the taper, and to depend on the taper of the draft of the pattern to give the finish.

1/32-in. depth of cut is a heavy cut when turning a piece 1½ in.in diameter
1/16-in depth of cut is as heavy as an 18 in. lathe will pull economically.
3/32-in. depth of cut is a good cut in turning a piece 10 in. in diameter on a 20-in. lathe.
1/8-in. depth of cut with 1/9-in. feed is a heavy cut on a cast-iron bar 8 in. in diameter.

For Boring Cylinders.

Allow 1/8-in. finish on each side of a 2-in. to 8-in. bore.
3/32-in. finish on each side of an 8-in. to 11-in. bore.
1/4-in. finish on each side of an 11-in. or larger bore.

Allow for one roughing cut and one finishing cut on all cylinders up to 11 inches in diameter. Allow for two roughing cuts and one finishing cut on cylinders 11 inches in diameter and larger.

In the foundry have the cylinder core made in a full cast-iron core box which has been bored out perfectly true and to size. Have the cores dried in half dryers, made exactly the same as the core box. Have the cores a tight fit in the core prints of the molds.

Finish on ends of pattern where mold parting crosses, such as the ends of a steam cylinder. Have the surface beveled from the edge of the flange to the bore of the cylinder. Allow no finish at the outside edge, and depend on the heavy bevel of the draft to give the finish.

Offset Shoulder.—A small offset cast on a piece that corresponds to an offset that is to be machined on the piece should be made a bevel on the pattern and not a square edge. See Fig. 13. Leave no finish at the very corner of the offset. This bevel will prevent the sand burning into the corner, and will save the cutting tool, thus speeding up the work.

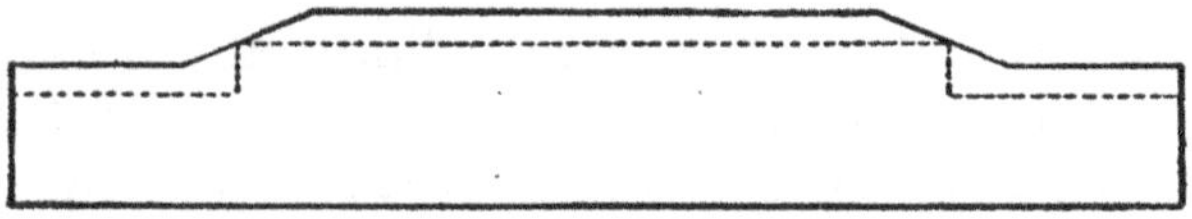

Fig. 13—Casting for an Offset Shoulder

Design Details to Save Machine Work

The designer will save the machine shop much work and save much money for his firm, if he will observe a few details which make it easier for the machine shop to accomplish its end of the job.

Use through holes wherever possible. These are more quickly drilled, as no attention has to be paid to depth. Where the hole is tapped, one tap can be run through instead of using two or three as with a closed bottom hole. A through hole is easier to stud or to put a cap screw into as no attention need be paid to cleaning out the chips in the bottom of the hole. Where tapped holes bottom, design them twice as deep as the diameter of the tap in order to take care of the chips and the taper end on the first tap. A 3/4-inch tap is ground back 5/16 inch.

The stud is run into the hole $\frac{7}{8}$ inch. There should be an extra allowance at the bottom for chips of 5/16 inch making a total depth of $1\frac{1}{2}$ inches or twice the $\frac{3}{4}$-inch diameter.

The following table of cast clearance holes will be found useful:

Diameter of Cast Clearance Holes for Studs.

$\frac{3}{8}$-in. stud, 15/32-in. hole at smallest point.
$\frac{1}{2}$-in. stud, 19/32-in. hole at smallest point.
$\frac{5}{8}$-in. stud, 3/4-in. hole at smallest point.
$\frac{7}{8}$-in. stud, 1-in. hole at smallest point.

Diameter of Cast Clearance Holes for Cap Screws.

$\frac{3}{8}$-in. cap screw, 15/32-in. hole at smallest point.
$\frac{1}{2}$-in. cap screw, 19/32-in. hole at smallest point.
$\frac{5}{8}$-in. cap screw, 23/32-in. hole at smallest point.
$\frac{3}{4}$-in. cap screw, 27/32-in. hole at smallest point.
$\frac{7}{8}$-in. cap screw, 31/32-in. hole at smallest point.

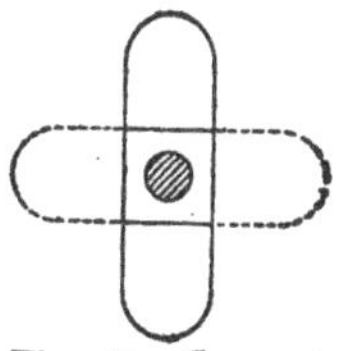

Fig. 14—Suggested Arrangement of Pieces to Give Flexibility

Great flexibility of location of one piece on another can be had by having the through bolts that hold the pieces together pass through slots in one piece that extend at right angles to the slots in the other piece, as shown in Fig. 14.

Designing to Save Time in Drilling

The increase in drilling sped accomplished by not having to change drills will be considerable, so design pieces with the same size of holes and same size of tapped holes all over. Avoid designing holes smaller than can be made with $\frac{3}{8}$-inch drills. This is the most economical size, as drills smaller than $\frac{3}{8}$-inch feed slower, break oftener, choke up more easily. The choking up of drills by the chips is especially bad when jigs are used. The chips will not come out through the jig bushings, and the jigs can only be used to start the drill. The pieces must be taken out of the jig in order to finish the drilling. To avoid this trouble a great deal of deep jig drilling nowadays is done from below so that the chips will pass out freely.

The cost of making drilling jigs can be reduced if a few standard lay-outs are used. The drafting room should keep a table of all the different sizes of drilling templates that are used in the shop, and should design so as to utilize these wherever possible. Cast depressions or drill spottings, wherever possible, to locate the drill. Work can be drilled faster by using this method of locating the holes than by any jig methods. It is better than laying out the work, as the one laying out on the pattern does for all time. It eliminates chucking time. It also has the advantage of being everlasting. A jig will wear out of true in a short time.

Designing to Save Lathe Time

Design turned corners with a groove in them so as they can be machined with the roughing tool, thus saving the time of putting in a square tool to finish the corner. See Fig. 15.

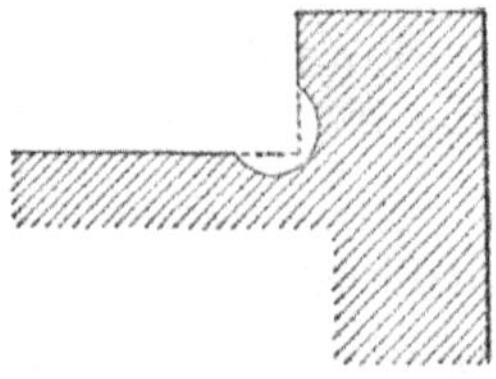

Fig. 15—Design of a Casting for a Turned Corner

If an arbor must be used, make it short. A 1-inch arbor 12 inches long will spring before the work slips. Where the end of a piece that is turned on an arbor has to be faced off, have a relieved part cast in the end at the center. This will save changing tools to machine the corner next to the arbor.

Study Weaknesses Which Develop

Keep a list of repairs in order to find the weak points of your product. Strengthen the design of parts that wear or break. Breaks in machinery are sometimes caused not by weakness but by carelessness, or by some peculiar condition or strain that the particular machine is under. The cause should be found and the machine should be changed to make it foolproof. If this is impossible, at least the reason of the trouble is known and customers can be warned. This may be knowledge which all manufacturers may not possess, and the one possessing it is then that much ahead of the rest.

A designer should have practical experience in the shop. He should not be afraid to discuss the points of his designs with foremen. He is not belittled by asking their opinions. Without their advice he may go far astray. For instance, consider the case of two rods to be held together by a long screw coupling. Suppose these rods must be roughly in line with each other. The design may be as shown in Fig. 16, Fig. 17 or Fig. 18.

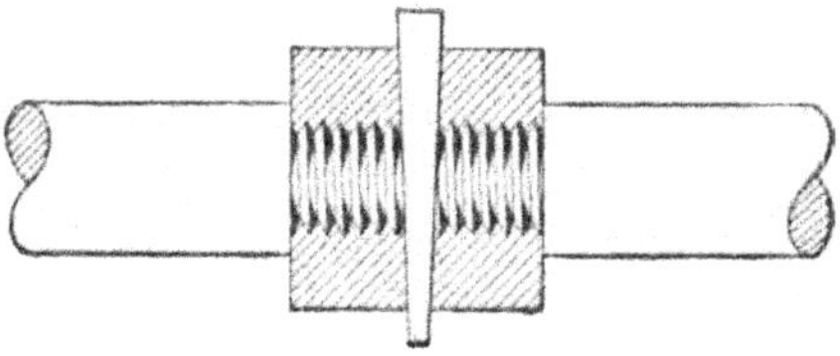

Fig. 16

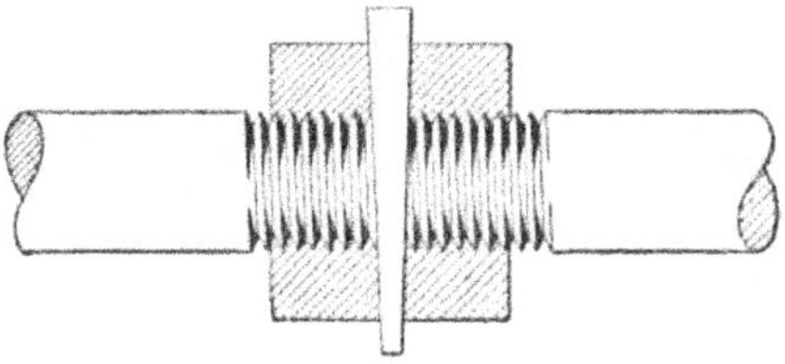

Fig. 17

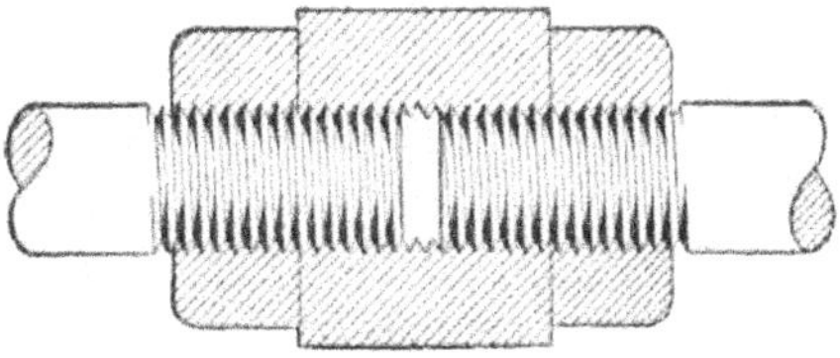

Fig. 18

Three Ways of Designing a Rod Coupling

The design in Fig. 16 first necessitates facing off the ends of the coupling. It requires the turning of the rod down to a shoulder which is expensive. It also requires threading in the engine lathe, as the thread must run up close to the shoulder. In doing this a great risk is run of getting a badly fitting thread.

The design shown in Fig. 17 looks fine on paper. The thread can be cut with a die. It is bad, though, because a facing operation is required on the ends of the coupling and the assembling is made expensive since the rods will not be in line with each other on account of being threaded with a die and on account of the threads in the nuts not being true with the nut faces. These inaccuracies will require repeated scrapings of the nut faces to bring things right.

The design shown in Fig. 18 is the cheapest way. Thread the rods with a die. Screw them into the coupling until they butt together. Test the rods to see which way they are out of line. Drop the rods and coupling on some solid object so that only the coupling strikes. This will bend the rods slightly inside the coupling and line them up. Drill pin hole and ream where the rods butt together and then drive in a pin.

The design shown in Fig. 17 will not straighten so readily by dropping, because it will be more out of line and stiffer than that shown in Fig. 18. Any bending will have to come outside of the nuts where the rods are rigid. This illustration is given to show that a designer may not always select the best way. The one which on paper looks the cheapest may not prove so in practice. Every part of a design may have points like the above, hence it should be discussed with the pattern shop, foundry, machine and erecting shops.

Some Fundamentals to be Observed in Design

The little details of design are the places where money is made or lost. If the designer adopts the standards given below he will not be very far from the right track.

A complete box shape is thirteen times more rigid against torsion, and four times more rigid against bending than the same amount of material in the form of side plates and thin cross girds. A casting in the form of three sides of a box has one-tenth the strength of a complete four-sided casting. Note the strength of a pasteboard tube. Slit this tube from end to end and its strengh is gone. The same is true of cast-iron sections.

Make the edges of flanges broad even if it takes a little extra metal. This gives a massive appearance to your machine often at little added but justifiable expense.

Use finished steel for all pins, rods, etc. Avoid turning of shoulders. A cheap way to make a shoulder, where there is no end thrust, is to slip a washer on the rod and hold it with a slit cotter. It is surprising in how many places a straight pin held with a couple of set screws could be used where the draftsman has designed with shoulders or taper ends. A straight pin, if practical, is cheaper than a taper pin. The reaming of a taper hole is a slow process and requires a new reamer.

Avoid small studs, as they are easily twisted off by the heavy handed.

Use cap screws with nut-sized heads where necessary. This will allow the casting or drilling of large clearance holes.

Screw Threads

Even though it be an expensive change, necessitating the scrapping of many good taps and dies, adopt the U. S. standard thread. Its advantages are: Taps and dies bought from different makers will be of exactly the same size. V-thread taps made by different tap makers differ. There is no exact standard for the V-threads. Another advantage of the U. S. standard is it has a clearance at the top and bottom of the thread; that is, the very pointed edge of the threads on a bolt does not bear on the bottom groove of the thread of the tapped hole. This allows a leeway for wear of dies and taps, and makes the U. S. standard die or tap wear twice as long as a V-thread. The objection to the V-thread is that when the taps and dies are worn they will produce a threaded hole in which a bolt that is apparently a tight fit actually bears only at the points and not on the sides of the threads. Such a bolt will soon become loose.

Adopt a few standard sizes of holes, taps, pins, rods, bolts, etc., rather than many; even though they seem out of proportion, for in this way a smaller number of gauges will be required, few sizes of material need be kept in stock, the tool equipment will be reduced and a smaller toolroom will be needed.

Use large clearance holes around cap screws and studs wherever possible to reduce the drilling and assembling costs. More time is spent accurately locating a drill than in drilling the hole. Large clearance holes will allow the use of badly worn jigs and will save the handwork in drawing over holes on the assembling floor.

Make all oil holes 3/8 inch. Use 3/8-inch split cotters.

Never design so that a piece has to be threaded up close to a shoulder for this makes it impossible to thread with a die, and lathe threading is rarely accurate and is always expensive.

Clearance Around Nuts

The distance from the center of the stud to the nearest wall should be as in the table below. See sketch, Fig. 19.

Size of Stud, In.	Dimension A, Fig. 19 Center of Hole to Face of Wall, In.
3/8	9/16
1/2	3/4
5/8	7/8
3/4	1 1/16
7/8	1 3/16
1	1 5/16

The values in the table are the minimum distances. Lower values will make it necessary to back face under the nuts and from 4 to 8 holes can be drilled in the time it takes to back face one bolt hole. Hence the importance of keeping nuts well away from the walls of the casting, especially if there is any considerable fillet in the corner or if the bolt circle is not concentric with the wall.

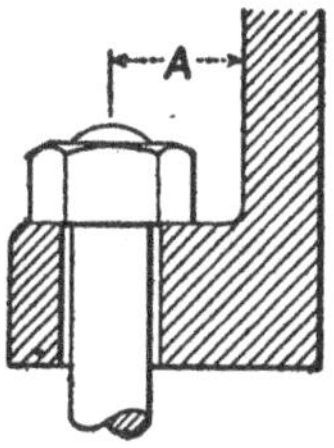

Fig. 19—Clearance Distance for Nuts

The designer can cause a great deal of confusion in the shop by giving names to pieces which do not fit. A simple rule for naming pieces is to call them by the two parts they connect. For instance, a lever link pin would be the pin that connects the lever and a link together. With this style of naming, any one can pick out the piece that is being talked about.

ARTICLE IX

SOME ECONOMIES IN THE FOUNDRY

Methods of Making Mixtures and Managing Cupolas that will Bring about Material Reductions in Cost and Increases in Output

THERE is a great saving in making the iron mixture of a foundry according to analysis. A well managed foundry's labor cost may run somewhere around 1¼ cents per pound of castings turned out. All the material bought by this foundry may run a little over 1 cent for each pound of castings. A saving in the cost of material by using two-thirds or more of off-grade iron or scrap will reduce the cost of the castings as much as increasing the foundry output per man 10 per cent. Yet this foundry will try to strain the output per man up 10 per cent. and pay no attention to the heavy expense of running a mixture half of which is costly pig iron.

A foundry using a minimum amount of coke in the cupola, so that the increase in sulphur from melting is only nominal, will be forced to use two-thirds cheap pig iron or scrap. To obtain a mixture in the casting of

1.80 silicon,
0.80 manganese,
0.60 phosphorus, and
0.10 sulphur,

more than two-thirds off-grade iron will have to be used. The above analysis will make a fine iron, close-grained, strong and soft.

A foundry that had never mixed by analysis used about 40 to 50 per cent. scrap in the mixture. This foundry after it changed over to mixing by analysis ran four months averaging 74½ per cent. scrap, or off-grade iron, at scrap-iron prices in the mixture. Three weeks of this time the mixture was 80 per cent. off-grade iron, three weeks 77 per cent. off-grade iron, three weeks 73 per cent. off-grade iron, and three weeks 69 per cent. off-grade iron.

The analysis of the iron ranged through this time:

Silicon 1 61 to 1 99,
Manganese 0 69 to 1 24,
Phosphorus 0 32 to 0 68,
Sulphur, 0 050 to 0 118.

with the average,

Silicon 1 82,
Manganese 0 90,
Phosphorus 0.55,
Sulphur 0.094.

It is strange that all foundries do not make their mixtures by analysis. The essentials of the subject can be mastered in an hour's time. The condensed statement of iron mixing can be completely told in a comparatively few words.

Permutations of the Five Metalloids

There are only five elements that affect the mixture: Three of these change during melting; the other two do not. Silicon and manganese reduce in amount and sulphur increases. It is a game played with a pack of five cards and it is a question of what combination can be made with five elements.

When an element is present in an iron above the normal amount, it changes the quality of the iron in a fixed direction, provided the other elements are at the normal point.

Silicon high, other elements in casting average: Soft, fluid, spongy, weak. When in excess of 3.50 iron begins to harden.

Silicon low, other elements average: Hard, strong. Less internal sponginess.

Manganese high, other elements average: Soft, strong, clean, close-grained. Manganese is a scrap carrier. When very high, manganese begins to harden iron.

Manganese low, other elements average: Hard, weak, brittle castings.

Total carbon high, other elements average: The mixture can carry a lower silicon with good results.

Total carbon low, other elements average: Strong iron. Increases the length of shrinkage.

Graphitic carbon high, other elements average: Soft, weak with internal sponginess.

Graphitic carbon low, other elements average: Strong, hard, no sponginess.

Combined carbon high, other elements average: Strong, hard, not spongy.

Combined carbon low, other elements average: Soft, weak and spongy.

Phosphorus high, other elements average: Fluid, weak and brittle. Pit holes on machined face if very high.

Phosphorus low, other elements average: Strong castings; molten iron sluggish in flowing.

Sulphur high, other elements average: Strong and hard, and sluggish in flowing.

Sulphur low, other elements average: Soft, but weak.

The proportion of the graphitic carbon to the combined carbon is purely a result of the proportion of the other elements, also the mold conditions.

Silicon and manganese high, other elements average, in a mixture with silicon 3 per cent. and manganese 1 to 1.50 per cent. The silicon softens and weakens. The manganese offsets this and strengthens.

Silicon and manganese low, the other elements average: Hard, weak, brittle.

Phosphorus and sulphur high, other elements average: Weak, brittle castings.

Phosphorus and sulphur low, other elements average: Strong and soft.

Phosphorus high, manganese low, other elements average: Hard and weak.

Silicon

	Silicon.
Malleable iron	0.60 to 1.00
Car wheels	0.60 to 0.80
Cylinders	1.25 to 1.75
Machine castings, heavy	1.75 to 2.25
Machine castings, light	2.25 to 2.75
Ornamental castings	2.40 to 2.75
Stove plate	2.50 to 3.00

On heavy work 1.60 silicon will be less spongy than 1.90 silicon.

For heavy pressure oil cylinders (hydraulic pressure) bring the silicon down to 1 per cent. to make a dense iron.

Five per cent. of the silicon burns out in melting in the average cupola. This has to be figured when making a cupola mixture.

Castings made 40 or 50 years ago, such as cast-iron building fronts, contain a higher percentage of silicon than the present day No. 1 iron. Such old iron is snapped up by the foundries that know this. It means getting No. 1 iron at scrap iron prices.

Manganese

Of all the elements this is the most desirable. It has no bad qualities and all the good qualities.

	Manganese.
Malleable iron	0.25 to 0.35
Car wheel iron	0.45 to 0.60
Cylinders	0.60 to 0.90
Machine castings, heavy	0.30 to 0.50
Machine castings, light	0.30 to 0.50
Ornamental	0.35 to 0.50
Stove plate	0.35 to 0 50

Manganese cleans, softens, closes the grain and offsets sulphur effects. High manganese removes sulphur. Often the sulphur does not increase in melting more than 0.002, if the manganese is 0.90 in the casting. This shows how manganese removes sulphur.

High manganese with high sulphur will make a softer mixture than low manganese with low sulphur. This shows the power of manganese as a softener.

Manganese is a better softener than silicon, as it does not weaken the iron nor make it spongy. It is cheaper, as high manganese iron is sold at about standard price, whereas high silicon iron is sold at a premium.

For average work 0.80 per cent. manganese is probably the ideal point. As the manganese is increased from a low point up to 0.60, its softening effect is felt very strongly. Above 0.60 the effect is felt less and less as manganese is increased, until somewhere about 1 per cent. any increase fails to soften the iron.

Castings will be hard if the manganese runs as high or higher than the silicon. This rarely occurs as such an analysis is unusual.

Makers of automobile cylinders carry 1 per cent. manganese in the iron to keep the mixture soft and the silicon low, to prevent sponginess in the unequal sections of the casting.

From $7\frac{1}{2}$ to 20 per cent. of the manganese that is in the mixture charged into the cupola will be burned out in melting, depending on the cupola. In making up a mixture, allowance must be made for this.

A good way to raise the manganese is to buy manganese scrap and use it in the cupola.

Carbon Content

	Total Carbon.
Machinery castings	3 50 per cent. and over
Ornamental	3 50 per cent. and over
Car wheel	3.50 per cent. and over
Cylinders	3 35 to 3.50 per cent.
Stove plate	3 30 per cent. and over
Malleable	2 00 to 2.25 per cent.

High total carbon makes a more spongy iron than low, other elements being average in amount. The higher the total carbon the softer the iron. Total carbon 3.75 per cent. and 1.50 silicon give a softer iron than 3.25 total carbon and 2.50 silicon. This shows the softening power of carbon. Low total carbon will produce excessive shortening of a casting and may crack it.

Northern and Southern irons differ principally in their total carbon and phosphorus. Northern iron has high total carbon and low phosphorus. Southern iron is low in total carbon and high in phosphorus.

Castings that cool quickly, either from being thin section, or from having a chill set in the mold, or from being poured with cold iron will be higher in combined carbon than those which cool slowly. This is the reason that a mixture that gives 0.50 combined carbon in castings $\frac{1}{2}$ inch thick will give 0.10 and 0.05 combined carbon in casting $1\frac{1}{2}$ inches thick. A mixture with 2 per cent. silicon in heavy work, and combined carbon 0.05 or 0.10 will change, by lowering the silicon to 1.50, to 0.50 combined carbon. Any sponginess found in castings made of the first mixture would disappear by using the second mixture. This shows that the proportion of combined carbon to graphitic carbon is a result of the proportions of the other elements.

A foundry troubled with internal sponginess will be obliged to use an iron mixture as near the hard point as the machine shop can stand. The silicon may have to be reduced to 1.50 per cent.

A cupola burns out the same fixed per cent. of silicon and manganese and adds the same fixed per cent. of sulphur to the iron day by day.

The mixture in the casting can be guaranteed except as to the carbons, knowing the analysis of the iron charged into the cupola, when the amount of change the cupola makes in the elements is learned from observation.

Phosphorus

From 0.8 to 1 per cent. phosphorus is desirable for fluidity in light castings.

High phosphorus weakens iron to a marked degree, especially if the manganese is low.

High phosphorus in medium and heavy castings often causes dirt.

Increase of phosphorus from a low point up to 0.6 per cent. makes a rapid increase in the fluidity, but not so rapid a change in weakness in the iron.

From 0.6 per cent. up the fluidity change is not so rapid as the change in weakness. For average work 0.6 per cent. is probably the most desirable point. This might be raised to 0.7 if the iron has to be carried a long distance or the cupola is liable to run a little cold. The loss of strength will only be slight and the increase in fluidity may save many castings. Carrying the phosphorus a little above normal will make the iron fluid and offset the sluggishness caused by high sulphur. High manganese being a strengthener will offset the weakening effect of the phosphorus.

Sulphur

The sulphur increase in melting iron comes from the sulphur in the coke.

A cupola running cold increases the sulphur more than a cupola running hot.

Of the sulphur in the coke 6 per cent. passes into the molten iron. This must be figured in making a mixture.

The sulphur in castings can be as high as 0.13 without any bad effect, provided the manganese is kept high enough to offset the sulphur effects. This is a valuable thing to know, as it opens up a field for melting off-grades of iron and scrap that will reduce the cost of the mixture.

High sulphur reduces sponginess in the heavy sections of castings. It produces sluggishness in running, which must be offset by increasing the phosphorus. The weakening effect from the increase of phosphorus will more than be offset by the strengthening properties of high sulphur.

Scrap Iron and Scrap Steel

A mixture of 66 per cent. scrap can be safely run continuously. I know of a hollow ware foundry that runs 75 per cent. scrap, also a pumping machinery firm that ran 75 per cent. scrap. These are two extremes on the casting line.

In certain markets it is cheaper to buy off grades of pig iron than scrap. A very high proportion of this pig iron can be used.

Analysis of average scrap cast iron:

Silicon	2.00
Manganese	0.35
Phosphorus	0.70
Sulphur	0.10
Total carbon	3.50

We will call the above an average mixture in a casting. Average analysis of scrap steel:

Silicon	None
Manganese	0.50
Phosphorus	0.08
Sulphur	0.08
Total carbon	0 50

Steel scrap is very useful in a mixture to reduce the silicon. Suppose you wish to run a mixture of two-thirds scrap and the iron in the casting comes out 2.10 in silicon and this makes a mixture that is too spongy and weak; add steel scrap to the mixture to bring the silicon down to 1.90, where you want it.

If a foundry can get the low silicon cheaper by buying off-grade iron instead of using steel, it is better to buy the off-grade iron.

Changes That Occur in the Cupola

Of the silicon, 5 per cent. in the mixture burns out.

Of the manganese, $7\frac{1}{2}$ to 20 per cent. burns out.

Of the sulphur in the coke, 6 per cent. is added to the mixture.

Suppose the silicon in the mixture in the cupola averages 2 per cent. Five per cent, of 2 is 0.10; 2 less 0.10 equals 1.90 per cent. The analysis of the castings would show 1.90 per cent silicon.

Suppose observation shows that the cupola burns out 20 per cent of the manganese, and suppose the iron charged in has an analysis of 0.75 per cent. manganese. Twenty per cent. of 0.75 is 0.15; 0.75 per cent. less 0.15 per cent. equals 0.60 per cent. The manganese in the castings will be 0.60 per cent.

If your coke has 0.75 sulphur in it, 6 per cent. of this would be 0.045. Now, if the iron that is put into the cupola, taking pig iron, scrap and all, averages 0.055 per cent. sulphur, 0.055 plus 0.045 gives 0.10. The mixture will have 0.10 per cent. sulphur.

Coke

Coke must be hard and in large lumps. If it is soft it will crush in the cupola and shut off the free passage of air and produce cold iron, and cold iron produces bad castings.

The higher the ash in coke the harder and the better; 8 to 12 per cent. ash is the best. Coke having lower ash, say, 5 to 8 per cent., is too weak to hold its burden.

Good coke runs 0.85 per cent. sulphur and under; the lower the better.

Taking Samples for Analysis

Casting Analysis.—Take eight gates from different parts of the heat. Drill them being careful to throw away the first chips that might have sand in them. Mix the chips from the eight gates to get a true sample. Put them into an envelope and send to a thoroughly reliable chemist. Better choose one whose principal business is doing this work for foundries, as the slightest error made by the chemist will lead to trouble.

Foreign Scrap Analysis.—Take chips from ten samples.

Coke Analysis.—Take chips from twenty pieces.

Here is the analysis of the castings of a firm which mixed satisfactorily by the old method of not using analysis. Its product was light castings:

Silicon	1.730
Manganese	0.660
Graphite carbon	2.900
Combined carbon	0.490
Phosphorus	0.944
Sulphur	0.058

They had no idea the sulphur was so low. They might just as well have used a cheap off-grade iron high in sulphur which would have raised the sulphur up to 0.10 or even 0.12 per cent.

Such an increase in the sulphur would necessitate the increasing of the silicon probably to 2 or 2.25 per cent., depending on how light the castings were, so as to offset the hardening effects of the higher sulphur. This change in the mixture would have reduced the cost of castings.

In starting up a cupola, light fire after two-thirds of the bed coke is charged. Light at tuyeres, not at tap hole or the bed will burn irregularly. After lighting, charge in the other third coke.

Begin charging the cupola as soon as the coke is thoroughly red hot.

Have the cupola fully charged 30 minutes before the blast goes on—not sooner and not later. In this way the full benefit is derived from the heat of the fuel.

Remember, the hottest place is next to the wall of the cupola; therefore, put the coke and pig iron there only.

Tap out when the blast has been on 30 minutes.

Never let the taphole blow, as it allows slag to run into the ladle, and slag will finally get into the castings. Always carry a large body of melted iron in cupola. The tapped iron then will come from the bottom of the molten mass where it is absolutely clean.

Be sure the ladle is heated before the first tap. Make the taphole 7/8 inch in diameter. The small taphole makes it easy to handle the iron under pressure in the cupola.

Close taphole with molding sand before blast goes on. This will preserve the bed coke and also will save some iron that is usually wasted.

Slag out after 50 minutes or after 7000 pounds on heats 12,000 pounds or over. Slagging will prevent the cupola closing up, thereby giving hot iron at the end of large heats.

Metal and Fuel Charges

Make the first charge of iron double the weight of those that are to follow. The coke in the bed is sufficient to melt it. Reduce the coke charges day after day until cold iron begins to result; then increase them a little.

Use steel scrap on first charge and pour the heavy castings with this iron. Steel is necessary in heavy castings to make them solid, because the silicon must be brought down low. All shop steel scrap melted in the cupola is clear gain for the foundry—material for nothing.

If you wish to melt cast-iron borings put 100 pounds in a sack and charge in the center of the cupola under the scrap—borings on the coke, scrap on top of the borings. Start in with 5 per cent. borings and gradually increase this day by day until 10 per cent. is reached. This material is no expense to the foundry and is clear gain.

Charge the coke, if it is good, to 24 inches above top of tuyeres.

One pound of coke in bed should melt 3 pounds of iron.

Use the least amount of coke possible. Excessive coke produces slow melting, raises the sulphur in the castings, and increases the material expense of running a foundry.

Coke that looks like the inside of a wasp's nest is not good. It makes slag.

Flux is not generally needed. Limestone is better than fluorspar. Fluorspar makes the flux so thin that it will flow with the iron and get into the mold. Use 30 pounds of limestone to 1200 pounds of iron.

Causes of Cold Iron

Cold iron raises the sulphur in castings.

Cold iron produces blow holes in machined faces.

Cold iron makes shrink holes under the risers.

Cold iron from cupola is caused by:

- Bed coke pieces being too small.
- Wood in bed not dry.
- Wood in bed longer than 2 feet.
- Wood in bed not level.
- Coke in bed and in charges not level.

Coke in charges when not put around the edge, but in the center, will make cold iron. Fork the coke in; do not pour it in.

Pig iron must be charged around the edge where the blast hits it, or cold iron will result.

Cold iron is produced from wet sand put in the cupola bottom.

Cold iron is produced when scrap is used in too large pieces. This cuts off the draft in the cupola.

Cold iron is produced by not charging scrap in the center of the cupola.

Cold iron is made by not charging the iron level.

Cold iron will result if the ladle is not heated before the first tap.

Cold iron is produced if coke is lit too early.

Cold iron is produced if iron begins to melt before blast goes on.

Distribute the shot iron throughout the charge evenly; not all in one charge or it will act as a damper and produce cold iron.

Hot Iron

Hot iron feeds the casting better giving less sponginess in casting.

Hot iron makes the casting solid under risers.

Hot iron makes clean castings.

Hot iron makes soft casting and castings free from blow holes.

Hot iron reduces sulphur.

Dry bottom sand gives clean, hot iron.

Two-foot pieces of wood or shorter give hot iron.

Dry wood leveled gives hot iron.

Use large coke in bed. A level bed gives hot iron.

Use small pieces of iron. If pigs are broken, especially for first charge, it will make hotter iron and reduce the coke.

Scrap in the center gives hot iron.

Scrap not too large gives hot iron.

General Directions

At the end of the heat the iron that is left over should be poured into a clean cast-iron pig bed, rather than on the floor. This will save the labor of breaking up the slab iron, put it into better melting form and keep it clean.

Remember, most of the dirt that is thrown into the cupola comes out of the tap hole, and some of it goes into the castings.

Be sure to oil the surface of the cast-iron pig chill to prevent the flying of iron.

Blast pressure should be 8 to 14 ounces.

Clay only 3/4 inch thick on the wall of the cupola gives clean iron.

Stiff, dry clay for daubing gives clean iron.

Be sure to skim slag from ladles before starting to pour.

Make slag hole 2 inches in diameter, 1 1/2 inches in length.

Put slag hole 2 inches lower than lowest tuyere.

Take out everything not used each day from the cupola top room. The day's heat can then be stored upstairs in wagons.

Line the cupola with standard firebrick inserting a wedge brick between every second or third brick to form the curve. A tighter fit and a more durable lining may be made thus than with expensive curved bricks. A close fit is of the utmost importance. Cracks between bricks on the inside of the cupola occasion a rapid burning away of the lining, and if the bricks stand apart at the back they become loose as soon as the lining burns thin. Lay up the melting zone in two walls, an inner and outer. The inside wall can be easily repaired and buring through to the steel shell is prevented. Bricks will be completely consumed before new ones are required, which is an economy. High manganese burns out the lining faster than low, but its advantage far offsets this trouble and expense.

Analysis and Expert Advice

Reputable firms of analytical chemists make a business of taking care of foundries. They make contracts by the year for analyzing samples from heats, coke, pig iron, scrap iron, or anything else, and give advice as to the best iron mixtures, for a fixed yearly sum. Besides this, they send their field man, free of charge, to teach the cupola tender the best way to run the cupola and the correct sizes of charges of coke and iron, etc. The field men are old practical foundrymen who have come up from the ranks. Troubles can be unloaded on them to be corrected. The price they charge is only nominal. The savings they effect will pay for their services ten times over.

ARTICLE X

LOSS AND SAVING OF CASTINGS

What the Foundry Superintendent Needs to Observe in Handling Molten Metal—Defects Frequently Found in Castings and Their Causes

THERE are 37 different ways of losing a casting:

1. Cold iron.
2. Dirt in the molten iron that fails to come to the surface.
3. Poorly skimmed iron.
4. Iron too hard.
5. Gas in the iron.
6. Mold too wet.
7. Cores not perfectly dry.
8. Mold rammed too hard all over or too hard in a spot.
9. Core rammed too hard.
10. Mold rammed too soft.
11. Mold not properly vented.
12. Cores not properly vented.
13. Tearing up of the mold in drawing the pattern and not patching the sand back properly.
14. Runout from a bad fit of the cores in the prints.
15. Runout from iron passing through a vent in the cores.
16. Runout from a strain through the solid sand of a mold.
17. Runout from poorly made parting.
18. Crush from a poorly made parting.

19. Crush from crooked setting of the cores.
20. Crush from poor fit of cores in prints.
21. Shift caused by flask pins being loose.
22. Drop out because too few bars or sand hooks were used.
23. Drop out because flask was weak or loose at the corners or at the bars.
24. Drop out because sand was too weak or too dry.
25. Drop out from something striking the top of the jagger after the mold was closed.
26. Dirt in the pouring basin, gate or mold or loose dirt around the tops of the risers.
27. Cores not wired or rodded strongly enough.
28. The raising of a core from not anchoring it properly.
29. From the casting cracking because it was gated wrong or from wrong iron mixture or bad design.
30. From internal shrinkage caused by improper molding or by wrong iron mixture or by cold iron or by bad design.
31. Failure to run or cold shut from bad iron mixture, too cold iron, too hard ramming, too thin a design or bad gating.
32. Scabbing of mold.
33. Cutting of mold.
34. Scabbing of core.
35. Restricted passage of gas after it leaves the core.
36. Dirt from the washing of core blacking.
37. Dirt from the washing of mold blacking.

An experienced foundryman can tell why a casting was lost by examining the bad spots. The flaw in the casting speaks plainly to him.

Gas Bubbles in Castings

Gas bubbles in castings come from three sources. They come from the mold or cores; from the iron having been poured too cold; or from a gas that is in the molten metal itself.

The first trouble can be avoided by care in making the mold and cores; the second by following exactly the rules for running the cupola and by handling the iron rapidly after it is in the ladle. A remedy that will completely eliminate the gas that is in the molten iron has not yet been discovered. This gas is not always present. It will appear in some heats and in others will not be in the least apparent. This gas always shows up in the shape of small, deep, penetrating, worm-like holes right around the place where the pouring gate enters the casting; never anywhere else. It will only show on the first castings poured with each ladle of iron. The gas seems to pass out of the iron after it stands in the ladle.

This gas can be partly eliminated by running the iron through long gate grooves around the parting before it enters the casting. Most of the worm-like holes will then come in this long gate.

Another way is to place beside the piece that requires very particular care a small pattern of something about which one is not particular—a pattern for a foundry clamp, for instance, or a cast-iron wedge pattern. Gate all the metal through the unimportant casting. Most of the gas bubbles will be in the clamp or wedge and very little in the particular casting.

Another way some foundrymen eliminate the trouble is to let the ladle of iron stand until the gas has escaped. This is not a good method; the last castings poured with the ladle are liable to turn out bad from cold iron.

Gas from the mold or cores shows in the shape of a violent kicking of the iron out of the risers while pouring. It shows in the casting in the shape of a hollow spot that is sometimes at a distance from the seat of trouble. Break up the casting carefully and trace the hollow spot piece by piece down to its lowest point. It will generally lead to a spot on the core or the mold that was improperly vented.

Sometimes a runout will cause a hole in a casting that can easily be mistaken for a gas kick. The difference can generally be told by the appearance of the inside surface of the flaw. The flaw from gas will be smooth. The flaw from a runout will show a granular surface with tiny points throughout the shiny interior of the hole.

If the trouble from gas continues, make a mold and cut out all the sand from the center of the cope so that the inside of the mold with its cores, can be seen. Pour the mold and the point that is giving the trouble will be quickly discovered.

Hard ramming will cause a violent boiling of the iron and will give a casting a streaked, wrinkled surface where the iron could not lie quietly against the hard sand. Wet sand will have the same effect.

Cold Iron

More castings are thrown out on account of having been poured with too cold iron than from any other source.

For a brief period in any mold, no matter how softly rammed or well vented or dried, there is a violent passage of gas from the sand into the iron. If the iron is cold and solidifies while this is still going on, the machined faces will be covered with little pit holes or smooth bubble-like hollows. The best way for a foundry superintendent to convince himself, the foundry foreman, the molder and especially the cupola tender of this is to take a casting that has been poured cold, have it machined early the next morning and bring it back into the foundry as an exhibition. Do this early before anyone has forgotten the cold iron that the casting was poured with.

Iron must come from the cupola dazzling white hot and must be hurried to the molds and poured quickly. Too much iron in the ladle will cause the last molds to be poured cold. Come back to the cupola often. Pour the left over iron into pig beds and not back into the tapping ladle. Have the phosphorus in the mixture high enough to keep the iron fluid.

Internal Shrinkage and Sponginess

Internal shrinkage shows in the heaviest part of a casting in the shape of a hole or a spongy place and can often be corrected by turning the pattern over so that the shrinky spot comes in the drag and not in the cope. Sponginess is rarely found in the lowest part of a casting. It can be corrected by putting a chill on the mold at the spongy place or by changing the proportions of the pattern so that all the metal is of the same thickness.

Sponginess very often shows in the bottom of a pocket or angle of a casting, caused by a disturbance of the iron from gas. Venting down into these corners or changing the design so that the angle comes on the inside of the casting instead of the outside will correct the evil. A pocket on the inside of a casting rarely shows sponginess.

A change in the location of the gate and risers will often eliminate shrinkage.

Putting on higher or longer risers will do away with sponginess by increasing the pressure in the mold.

Venting down into corners that show leakage will often correct this evil, as any gas disturbance of the iron in the mold will make a porous spot in the casting at the point of disturbance.

Plates of unequal section are likely to crack. The trouble can be remedied by using a stronger mixture of iron or by using one with a higher total carbon or by changing the location of the gate so that the point that cracks will be the first to cool and not the last or by changing the design of the plate to one of equal section all over.

Gating and Pouring

Make a pouring pocket at the top of the gate into which the iron is poured. It will catch the first splash of iron and keep it from going down the gate.

Make a pocket at the bottom end of the gate to catch the first drops of iron and keep them from splashing into the mold. These first drops of iron chill in the mold and make bad spots in the machined faces on castings.

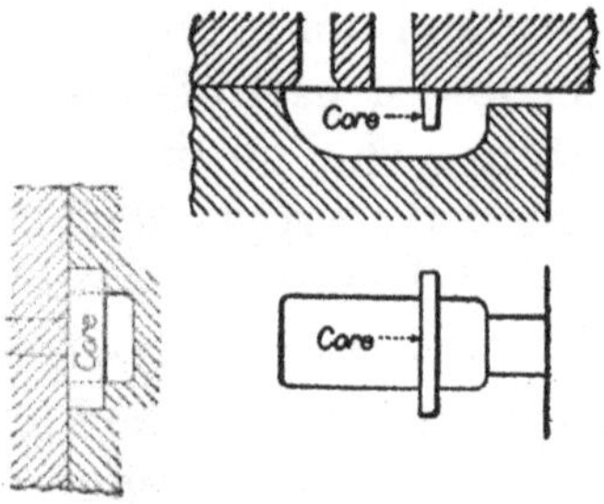

Fig. 20—Slag Skimming Gate for Molds

Soften up the gate at the pouring bowl also at the parting with a lifter before slicking these places down. This will prevent the boiling of the iron, which creates dirt.

When pouring be sure to keep the gate choked full of iron to prevent the slag getting into the casting.

Make a skim gate, using a pattern and a core in the manner shown in Fig. 20. Its advantages are: 1. The sand is soft in the drag. 2. The core acts as a perfect skimmer. 3. Using a pattern instead of cutting by hand makes a clean gate.

A rod for churning should extend into the casting.

A hot poured casting will be softer than a cold poured one of the same analysis for the sand, being heated to a higher degree, will anneal the casting.

Dirt in Castings

Dirt in castings can be caused from sand left in the mold before closing; dirt in the pouring basin at the top of the gate; loose dirt that has fallen in from the top of risers because the molder is careless about leaving the sands ragged at this point.

Dirt may come from a cutting action of the iron in the mold or from a scab on the core or mold or from a crush on the mold. It can come from the blacking washing.

When a casting shows up dirty examine the spot. A fine black powder is blacking.

Whether a casting was lost from a piece of broken core or a piece of the mold can be easily told if limestone is present in the core sand. The molding sand will be black, whereas the core sand will have white particles in it. A careful examination of the casting will show the spot from which the sand came. If it is at the gate the sand was washed away by the rush of iron into the mold, which can be prevented by driving nails over the inside of the mold at this point or by changing the gate.

Washing on heavy castings is frequently caused by using a grade of sand that is suitable for light work only. If the trouble is from a scab on the core it shows that in making the core the sand became loosened and never was pressed back again before drying.

Sometimes the slicking of a wet blacked core with the tool, if not properly done—that is, if the core-maker lets his tool lie too flat against the face while slicking—will loosen the core sand.

If the trouble is from a crush at the parting the molder failed to depress the parting properly by slicking.

A scab on other parts of a casting is caused by the sand buckling out into the casting. The molder rammed too close to the pattern in the spot where the scab occurred.

A poor grade of blacking may wash and make dirty castings.

Adulterated blacking is all right if you do your own adulterating. Buy the finest grade of plumbago at 5 cents per pound and mix it with as much talc, at 1 cent per pound, as will work satisfactorily. Thirty-five per cent. talc and 65 per cent. plumbago for a wet blacking mixture will peel the sand off the heavy castings. Half and half will be all right for the average blacking. Pure talc, with no plumbago, will be found perfectly satisfactory for dry blacking thin plate work.

Blacking often washes because not enough molasses was used to bind it. It washes if put on too thick or if burned in drying.

Put stone coal in the sand pile and mold blacking can be eliminated on certain classes of work.

ARTICLE XI

METHODS WHICH SAVE IN MOLDING

The Part Played by the Different Classes of Machines in the Foundry System with Directions for Operating Them Economically

MAKING a mold involves a great many small simple operations which have to be carefully done. The least neglect or lack of skill in doing any one of these many small operations means the loss of the casting. Molding is usually done by having one man carry the mold through all its steps from the start to finish. This requires a high-class, physically powerful man who has served a long time at the trade.

Each step in making a mold is simple and easily learned. Therefore, the tendency of the times is to have one man do one or two of the operations of mold making only and pass the mold on to others to do the following operations.

The jar ram molding machine fits into this system admirably. Under this system each man becomes more expert at his one or two simple operations of mold making than the best skilled mechanic. He saves time by keeping all his appliances for this one task right with him. His wage rate is lower than the rate of a fully skilled molder, because he is not paid for a complete knowledge of molding. Each of these men turns out more work than a molder would, because any slowing up makes the work accumulate. Each has to do his part of the molding as fast as the molds come to him, so as to get them off his hands and to the next man.

All these points in the system reduce the cost of castings materially below that of the old way. One foundryman in Cincinnati put it this way: "Nine-tenths of the steps in producing a mold are plain operations that a laborer can do. Our aim is to make laborers do all these and use the skilled man on the skilled tenth of the work only."

The system is worked out as follows: One man, or a small gang of men, temper sand all day; a second gang fill the flasks, jar them and deliver them to the finishing gang, who draw the patterns, tool the molds, if this is necessary, and black them. The next man or men dry them. Most of the molds are left open until the last thing in the evening and then all hands jump in and place the cores, close the molds and clamp them.

The gangs are kept down to as few men as possible. At blast time there is always a mold left open, not cored, another rammed with the pattern still in it. These molds are ready to start on the next day. The gang starts right in at whistle time turning out molds.

All the sand in the foundry is kept at the jarring machine. All the molds are taken to this point by the crane to be shaken out. The shaking out goes on all day, whenever the laborers get spare time. Each flask is placed on the follow board the instant it is shaken out. This saves extra handling.

The system of shaking out and cutting the sand as it is needed saves the expense of a night gang. It reduces the day labor gang to the minimum; uses the crane evenly all day and cuts out waiting on the part of the molders for the crane.

A mold that cost $3 to make by the old system of one man molding will be reduced to about $1.90 by the gang system using a jar ramming machine. This system is adaptable only to molds larger than 24 x 24 inches inside measurements.

Unit Output for Different Kinds of Molding

A foundry having intricate core work on about half the tonnage will find the output of castings per man connected with molding running about as follows:

400 lb. per man at old style floor work system.
550 lb. per man at bench work.
625 lb. to 830 lb. for small molding machine work.
650 lb. for gang system at jar ramming floor work.

These figures will run higher for foundries making less difficult work. The relation of pounds output per man on the different kinds of molding will stay about the same. The small molding machine will hold the record and the jar ram floor work will follow next.

A foundry using the gang system of molding will arrive at such a point of independence that the loss of the best molder in the shop is not felt very much. It is generally harder to replace the exceptionally good helper who has adjusted himself to the methods of the shop, knows where everything is kept, knows all the sizes of the flasks, brings out the cores and places them beside the molds, and fits himself into all the chinks of the foundry, than it is to replace a molder.

The Scope of Different Molding Methods

On all molds up to, say 14 x 16 inches inside measurement of flask, the squeezer molding machine will produce faster than if the molds are made on the bench or by the jar ramming machine.

On all molds 26 inches square and larger the plain jar ramming machine will produce faster than hand ramming.

Molds between the squeezer size and the jar ramming machine size are not as yet made economically by the molding machine. There are machines on which the molds are hand rammed. The drags are deposited on the floor by the machine. There are others that jar ram the molds, but do not deposit the drags on the floor. This size of work will not be satisfactorily done until a machine is made that rams the mold by the jar ramming process, deposits the drag on the floor and moves along, or is moved along, to the next position.

Any molding machine to be a success must be a very simple mechanism—the simpler the better.

Molds 14 x 16 inches inside measurement of flask and smaller should be made on a hand-squeezer rollover-pattern drawing machine that handles both cope and drag at one squeeze and draws the mold down and not up from the pattern. The pattern must be above the sand when being drawn, otherwise the pattern making expense will be greatly increased; only a very perfect pattern will draw down from a ceiling of sand without pulling the sand with it. A job with a hanging body of sand cannot be made by lifting the sand up off the pattern.

Using the Squeezer Molding Machine

Copes 8½ inches deep and drags 8 inches deep with the pattern extending 6 inches from parting, can be handled perfectly on the squeezer machine. The ramming will be done with the shovel handle on this deep work. Such a squeezer molding machine will hold the record tonnage output in a foundry. On small molding the greatest economy is made by reducing the motions that the man goes through in making a mold. This can be done on the machine that makes the cope and drag at the same time. The man picks up his shovel only once in making a complete mold. He fills the cope, drag and sieve with sand and rams the mold all at one handling of the shovel. He strikes off both the cope and the drag at one sweep; puts on both bottom board and squeezer board at one movement; a movement clamps both cope and drag; they are both squeezed with a single motion, and the patterns are drawn with a return motion of the handle; the mold is closed and carried out as a whole in a single trip.

On this kind of work the pattern board has the sprue and riser post of brass mounted on it. On the squeezer board is a form that makes the pouring bowl on top of the mold so that this handwork is eliminated.

It is best to adopt a layout for the gates and risers that will cover all cases and never vary from this. The best layout for the gates and risers is at each end on the center line, at each side on the center line, in each corner, two near the center crossways, two in near center lengthways, and one exactly in the center. Nearly any combination can be worked with a standard outfit of squeezer boards by using this layout.

Economies With the Squeezer Molding Machine

Place vents permanently at the parting on the pattern board to save cutting them in the mold by hand. A plain job in a 14 x 16 inch mold can be made every 2 minutes using bands in the flask, and in less time if solid flasks are used. This is timing the man on a single hour's run. He will not be able to keep up this rate all day, but it shows how rapidly molds can be made on the small machine.

The output on the machine depends entirely on the strength of the man. A very powerful man can put up 100 molds 14 x 16 inches between 7:00 a.m. and 11:00 a.m. He will have to be physically fit for the task—built on the lines of a heavy freight locomotive. It takes 19 minutes to change the boards, squeeze, etc., on such a machine.

Teach the man carefully to make no false moves; to peen with the shovel handle a certain number of strokes —no more or no less; to lay his tools always in the same place, and to remove the sand with one sweep, not two, when striking off the mold. A green laborer the third day he works, if carefully taught, will make a 14 x 16 inch mold at the rate of one every $3\frac{1}{4}$ minutes.

Laborers should shake out the small machine molds as soon as they are poured, pile up the castings, throw water on the sand and next morning temper the sand for the machine man before he starts. If he is forced to cut his own sand he will not be able to put up a big day's work. Some firms even go so far as to have a different gang pour off the molds from those that make them. A man can then go the limit all day without having to face the tiring task of carrying and pouring half a ton of molten iron at the hour when he is already worn out from molding.

In piecework foundries everything is in readiness for the molder to start in the morning, his sand is cut and the pattern is in his machine.

The Case of the Jarring Machine

Jar ramming is the only perfect way to ram molds larger than 24 x 24 inches. Any saving in time made by the jar machine on smaller molds is lost in the labor of placing the molds upon and taking them off the machine. It is safe to assume that the ramming of molds larger than 24 x 24 inches consumes 20 to 50 per cent. of a molder's time, depending on the style of the work. The jar ram machine, by abolishing hand ramming, will save 20 to 50 per cent. of the molding expense. An hour's ramming can be done in a minute; the ramming is perfectly done. The even ramming eliminates scabs and swollen spots on the castings.

The molds come out very smooth, that is, no tiny particles stick to the pattern as with hand ramming. There is a slight sliding of the sand on the surface of the pattern, or a slight give of the pattern that keeps all particles free from any sticking tendency. The inside of a jar rammed mold is as smooth as velvet. The sand is always hard and strong at the corner of the parting where the pattern meets the follow board. No filling in of sand at the parting after the mold is rolled over is necessary as with a hand-rammed mold. The sand is hardest next to the pattern and is softer back from the pattern so that venting is unnecessary. This saves time and saves marring the pattern with the vent wire. Preserving a smooth surface on the pattern saves the time of tooling the surface of the mold.

The jar ramming machine must not only be served by the main foundry crane, but must have a quick handling boom crane of its own. The output of the machine is controlled by the speed at which molds can be put on and taken off. One minute is all the time that is required to ram a mold, so that one machine well equipped with mold-handling apparatus will ram all the molds for a large foundry. A pattern drawing attachment on the jar ramming machine rarely pays, as the slight time saved in drawing pattern is more than lost by the time wasted in changing the machine from one pattern to another.

Working a Gang with the Jarring Machine

Following is the time of making a mold on the jarring machine by the gang system working at the regular speed that is kept up all day. Foundrymen can compare this time to their own mold-making time and see how much saving the gang jarring system would give them:

Drag 30x36 In. Inside Measurement, 18 In. Deep.

1 min.—Putting the pattern and the drag on the follow board.
1¾ min.—Clamping the flask and follow board together.
6¾ min.—Sifting sand around the pattern and shoveling in the sand.
2 min.—Sifting sand on top of the pattern.
2 min.—Shoveling the drag full of sand.
1 min.—Putting the sand frame on and filling it with sand.
½ min.—Crane placing drag on the jar ramming machine.
½ min.—Jar ramming.
1 min.—Putting bottom board on.
¾ min.—Clamping bottom board.
½ min.—Crane takes the drag to the molder to finish.
¾ min.—Take off follow board.

NOTE.—The time on all molds of a size will be about the same, no matter what the pattern be up to this point. This next item will vary with the style of the pattern although the time will be short with the gang system, as the men work rapidly.

10 min.—Finishing the drag up to the point of blacking.
½ min.—Spray the drag with blacking.
5 min.—Drying with an oil torch.

NOTE.—Some foundries diminish this labor cost by drying the molds in ovens, instead of drying with a torch, which will pay if the foundry has the crane capacity, the room, and a handy oven arrangement. The molds should go in at one end of the oven and come out at the other.

Cope 30x36 In., 6 In. Deep.

1 min.—Put pattern and cope on follow board.
3½ min.—Sifting sand upon the pattern, getting the gaggers and clay washing them.
2 min.—Setting gaggers.
2¼ min.—Filling cope with sand.
¾ min.—Putting on a sand frame 7 in. high and filling it with sand.
¼ min.—Crane taking cope to the machine.
½ min.—Ram cope.
¼ min.—Take cope off machine.
½ min.—Take off sand frame and shovel off the extra sand from top.
1 min.—Strike off cope.
½ min.—Crane taking cope off machine and turning it over.
1 min.—Setting it down at mold finisher.
½ min.—Take off follow board.

NOTE.—The following time item varies according to the job.

10 min.—Draw cope pattern and finish the mold.
1 min.—Spray wet blacking on the mold.

The rest of the mold making would run the same as ordinary molding when done at a rapid rate.

Wet Blacking of Molds

There is great economy in the wet blacking of molds. This can be done in one-fifth the time required for dry blacking. The total length of time, including the drying, will be less than that consumed in dry blacking with a camel's hair brush. Spray the blacking on with compressed air by an atomizer.

If the drying is done with a hand torch, mix the blacking thick and spray it on. Smooth the blacked mold with a camel's hair brush dipped in water. This will cut down the drying time. If the molds are oven dried use the blacking thin and no hand brushing will be needed.

A casting made in a wet blacked mold will come out clean. A single light blow of a hammer will knock off the sand, thus saving labor in the cleaning room. Castings free from sand reduce the machine shop time.

A dried mold has a hard, clean surface for the iron to lie against. The blacking hardens and cements all the loose corners to the mold.

Use the coal oil torch for drying. Make your own coil pipe for the torches when they wear out. One laborer will dry a great many molds in a day and will become very expert at it.

ARTICLE XII

PATTERNS FOR MOLDING MACHINES

Placing Patterns on the Small Molding Machine—The Sizes of Flasks and Other Important Details—Examples of Foundry Output

IN putting patterns on the machine mount an accurately machined cast-iron frame for the drag and one for the cope. These frames must be backed with machined cast-iron plates. The drag frame has holes at the ends to receive the pins of the drag and the cope frame has pins in it for the holes in the cope.

The patterns are mounted on wooden panels, a panel for the drag and one for the cope. These panels drop into the frames and are held in place by four small sliding plates or clamps in each frame. The clamps lock the panel boards down securely by screws. The boards are located or centered accurately in the frames by pins in the depressed metal backing which enter pin holes in the backs of the boards.

The advantage of this system over mounting the patterns on boards or plates fitted with flask pins and pin holes is that it simplifies pattern work. The flasks need to be fitted to but one set of frames. Great care can be used in doing this. The system where each board has to to be matched perfectly with the flask pin and pin holes requires an amount of continued accurate work, which has forced its abandonment in many foundries. Many castings will show shift on account of doctored up pin holes in pattern boards made according to the old way.

Material and storage space are saved by the panel system.

Split patterns are correctly located on the panel boards in the foundry as follows:

Procedure to Get Matching of Patterns

1. Make cope and drag boards to an exact size.

2. Screw to the back of the cope board a hardwood drilling jig, being absolutely sure two edges of drill jig match two edges of the cope board.

3. Drill the two dowel-pin plate holes, using bit on machine to drill square.

4. Do same with drag board.

5. Put brass dowel-pin holes in both boards.

Note: The molding machine is equipped with a frame that has dowel pins in it to match these holes.

6. Have cope and drag split patterns doweled together accurately in the regular way.

7. Place drag board into the molding machine frame.

8. Lay those halves of the patterns with pins in them on the drag board, first drilling loose clearance holes in the board to clear the pattern pins.

9: Ram up a drag. Ram very hard.

10. Turn the drag over, lift off the drag board leaving patterns stuck in the drag.

11. Set the cope half patterns upon the drag half patterns. Ram up the cope same as in regular molding and lift off the cope.

12. Put glue on the parting faces of all patterns.

13. Put the drag frame with its drag board, back upon the drag and clamp it tight against the patterns, making the pattern stick to the drag board.

14. Do the same with the cope frame and cope board. Let them stand over night until the glue has set.

15. Lift out and permanently brad or screw patterns into place.

Note: If the sand is cut away all over parting before glueing, a better contact is made. The whole operation is done in a few minutes. No measuring is required and the matching is perfect.

It is a good plan to have a vertical plate set on a surface plate in the pattern shop upon which the cope and drag panels can be clamped, one on each side, to test the accuracy of the pattern placing and to lay out panels for the locating of any patterns that cannot be placed by the foundry process. This vertical plate has dowel pins on each side that locate the drag panel and the cope panel exactly opposite each other accurately. The plate has feet on its edges so that it sets perfectly plumb or perpendicular on the surface plate.

A pair of panel boards are clamped in place on opposite sides of the vertical plate and a horizontal line is scratched with a surface gauge at the point where the drag pattern is to be placed. A corresponding line is scratched on the cope panel. The paneled plate is then turned edgeways one-fourth of a turn. A line is scratched again on both panels. This line intersects the first at right angles. The patterns are placed on the panels accurately with reference to these intersections.

Follow Boards for the Jarring Machine

Make a table of pattern pin distances for use as a standard, such as given in Fig. 21. Plan the distances so that a pattern that should be placed lengthwise cannot be placed crosswise. The method of locating the pin holes in the follow board is as follows:

1. Put pin holes A and B, Fig. 22, into board for the flask pins.
2. Draw a line through A and B.
3. Mark off point C midway between A and B.
4. Draw main center line DE through C. Note: This line DE must be at exact right angles to AB.

F and G are holes in the pattern board for pattern pins. The line FG must be parallel to AB. The distance from F to H must be exactly the same as G to H.

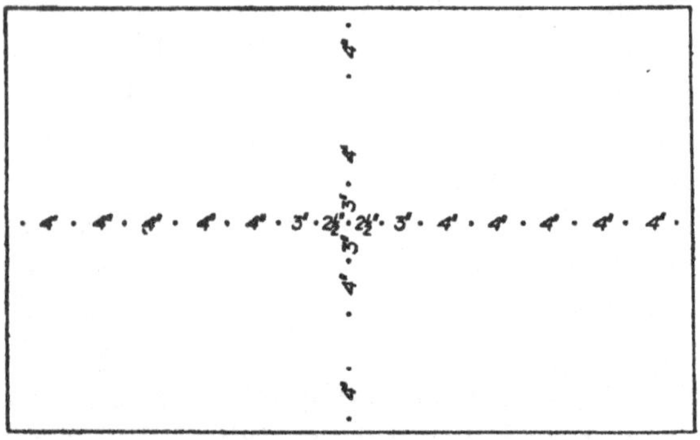

Fig. 21—Suggested Table for Pattern Pin Distances

It makes no difference where H comes on the line DE. It can come at C if necessary. In practice FG is generally located on the line AB. The pin holes then come on the line AB.

Placing Pins in Patterns

The procedure in placing pins in split patterns is as follows:

1. Place the pins in the cope pattern a distance apart equal to the distance from F to G on follow board. That is the distance between pins in the cope pattern must be such that they match the pin holes in the follow board.

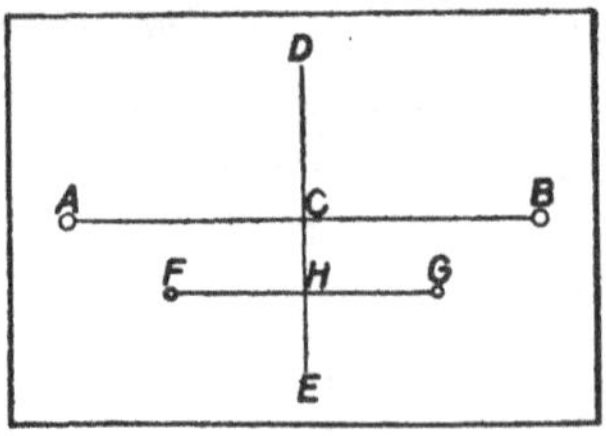

Fig. 22—Locating Pins in Follow Board

2. Pin holes in the drag pattern must match pins in the cope pattern and the drag pattern must match the cope pattern.

Note: It makes no difference where these pins and pin holes in the pattern come in relation to the pattern.

That is they can be out of parallel with center line of pattern. They can be nearer one end of the pattern than the other.

3. Exchange pin holes in the drag for pins.

The procedure in the case of *half patterns* is as follows:

1. The distance between pins in the pattern must match the distance between pin holes F and G in the board.

2. A line through these pins IO in Fig. 23 must be parallel with the center line NJ.

3. The distance from I to the cross center line LKM (that is IK) must be exactly the same as OK.

Note: It makes no difference where K comes on the line LM. It can come at the point where NJ crosses LM, if necessary.

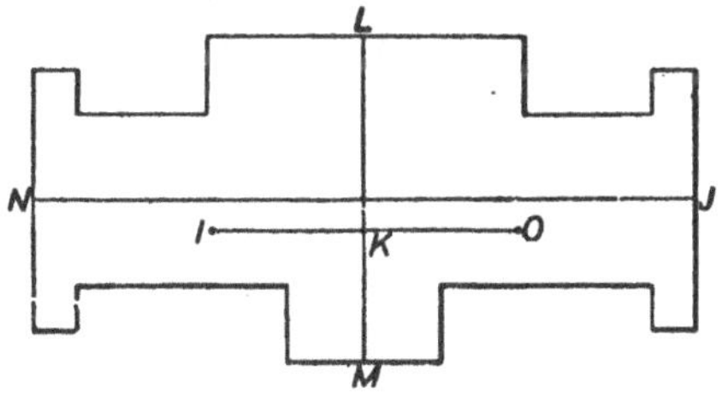

Fig. 23—Diagram for Locating Pins in Pattern

All the dimensions of the pins for the patterns and the pin holes for the follow boards are made to exact size. The flange around the bottom of the pin is perfectly true with the pin. The pin hole is exactly central in its piece of brass.

The process in detail of putting the pins into the patterns is as follows: Tram points are set accurately to a steel rule. The distance is marked on the drag pattern from the center K out to the correct location of pattern pins I and O. Circles are described at I and O the exact size of the pattern pin plates. With a very sharp tool the pattern is routed out inside these circles the exact diameter of the flanges on the pattern pins. Before placing the pins into these routed depressions, special tool steel disks are inserted having sharp raised edges on them that will describe a circle the exact diameter of the outside of the flanges of the pattern pins. The sharp edges stand above the surface of the drag pattern 1/16 to 1/8 inch. The cope half of the pattern is now laid on and the edges carefully matched. When it is perfectly set it is hit with a rubber maul. This jams the patterns together and the tool steel disks describe circles on the cope pattern, locating exactly the pins that go into this half.

The foreman patternmaker keeps a couple of hardwood blocks, each block having a hole drilled through it the exact size of the pattern pin. He drops these blocks over the drag pattern pins and places the cope pattern upon the drag pattern. The cope pins enter the holes in the blocks. The blocks locate the pins of the cope pattern opposite the pins on the drag pattern and hold them slightly apart from each other. When this is done he sights around the edges of the patterns and sees that there is no shift.

A pattern can be rammed on the jarring machine by fitting pins into the old pin holes if they be spaced correctly to match the follow board, holding the pins in place by wood screws running lengthwise through them into the wood at the bottom of the holes. In this way a pattern can be either hand or jar rammed.

When follow boards become worn where the flask rests, lay strips of old leather belts on the worn places. This will raise the flask enough to insure the mold joint being sand to sand and not flask to flask and prevent runouts.

Suggested Sizes of Flasks

In order to make a small equipment of flasks cover a wide range of work, plan their sizes and shapes according to a system. Three shapes will probably cover all your needs: a flask nearly square, one a little less than twice as long as it is broad, and one more than twice as long as it is broad. Adopt a system of spans for pins and make the flasks to suit them—something like this:

Size of flask, in.	Distance between pins, in.	Size of flask, in.	Distance between pins, in.
14 x27	30	30x 89	92
17½x17½	20	38x 41	44
18 x31½	34	38x 69	72
20 x35	38	38x113	116
20 x59	62	48x 53	56
24 x27	29	48x 83	86
24 x41	44	48x183	146
24 x71	74	60x 69	72
30 x35	38	60x105	108
30 x53	56	60x179	182

The pin distances on this list are 20, 30, 38, 44, 56, 72, 86 and 108 inches for the standard flasks and 62, 74, 92, 116, 146 and 182 inches for the long flasks. If the flasks are made of structural iron the pin distance of 30 inches ought to be reduced to 29 inches on account of the flange being narrow.

Good flask equipment pays. An inexperienced man can make molds with good equipment. Only the best molders can use poor flasks. Channel iron flasks are easily made. Saw a slit on the flanges at the points which are to become corners. Heat the channel in a blacksmith fire and bend. This makes a very stiff, durable flask with a riveted joint at one corner only.

Use sheet steel bottom boards with angle iron battens on the under side of flasks up to 30 x 35 inches. Above this, use wooden boards. Steel plates do not burn, but are not satisfactory in large sizes, as they are too heavy to handle, and they warp if there is a bad bottom runout.

Foundry Output

A man in a jobbing foundry makes molds for chain pump sprocket wheels at the rate of one every $3\frac{1}{2}$ minutes. He sets 10 cores in each mold.

I noticed in Cincinnati, on furnace work, an Italian molder had 105 molds up at 10:50 in the morning. The flasks were 10 x 12 inches, 5 inches depth of drag and 5 inches cope. There were six long narrow patterns in each mold.

The best two men in a jobbing foundry in Indianapolis put up 210 molds together in a day, one man working on drags and one on copes. The castings were washing machine wheels 24 inches in diameter; the copes were 4 inches deep and the drags the same depth.

On plain plate work one man will put up 50 floor molds, 24 x 24 inches and 10 inches deep by 4:00 p. m.

The molding machine companies claim 275 molds in 7 hours on 11 x 16 inch flasks, 1200 pounds of iron poured. They get this wonderful record by paying the man an enormous bonus and allowing him to rest up for a number of days after the record is made.

Jobbing foundries sell furnace work at $2.25 per 100 pounds, stove plate at $2.40 per 100 pounds, and the lightest work at $2.75 per 100 pounds, when pig iron is $17.50. Their general cost, which has to be added to the piece price given the molder, runs about $1.50 per 100 pounds on the prices named. Corliss engine castings, I am told, cost $1.70 per 100 pounds to make.

Foundry loss of castings runs about 10 per cent. on the average; well-run foundries about 5 per cent.

A Mixture for White Metal Patterns

Lead, 5 parts.
Tin, 3 parts.
Antimony, 1 part.

Bismuth, 2 parts.

Use all new sand for the molds and ram very hard. Pour the metal very cold. The metal should not be hot enough to ignite brown paper.

ARTICLE XIII

THE SUCCESSFUL MACHINE SHOP

How Avoidable Losses May Be Prevented—When to Buy a New Machine—Where Profits Are Injudiciously Sunk—The Night Shift

THE average machine shop operates at about 30 per cent. of its capacity. To test the truth of this statement, step out into the shop an dnotice how many machines are running one speed slower than they should. The tool should be cutting at the rate of 60 to 100 feet per minute for outside turning of steel and cast iron; at 40 to 60 feet per minute when boring on inside work, and at 60 to 100 feet per minute when milling.

Drills of high speed steel should be spinning at a fast rate. A 5/8-inch drill should be revolving so fast that you cannot see the flutes. The modern type of twist drill will drill through 15 inches of cast iron in a minute, or through 4 inches of steel per minute. Is it drilling that fast? The answer will be "No" every time. To show how the new tool steel has changed conditions, the highest speed on a radial drilling machine bought 8 years ago, is too slow for a 5/8-inch drill.

Are you trying to turn work on long, slim arbors? It is impossible to take a heavy cut on an arbor unless it is short and thick. Either the work will slip or the arbor will spring.

Look at your steel and cast-iron borings. Are all the steel borings blue in color from the heat generated by the high speed at which the machine was running? Are your cast-iron chips the size of grains of wheat and corn, showing that heavy cuts have been taken? Or are they a pile of iron dust, showing that fine feeds have been used? The feed on all machines should be so heavy that the cutter chatters on the roughing cut. Let the finishing cut take out the chatter marks.

Where to Look for Avoidable Losses

Are the jigs and fixtures such that it takes less than 2 minutes' shutdown of machine between the end of the finishing cut on one piece and the start of the first cut on the next piece?

Do you use drill spotting marks cast in the castings to locate the point of the drill, thus avoiding the use and expense of jigs and saving the expense of laying out the holes?

Examine the design of your product piece by piece. You will find by redesigning it that 20 to 30 per cent. of the labor cost can be eliminated.

What amount of finish is allowed by the pattern maker? Is it so heavy that the machinist is forced to take three cuts to bring the work to size? He should take but two cuts.

Watch your vise hands. What tools do they use the most, the wrench or the hammer, chisel or file? Parts should come to them from the machines so that erecting is merely a bolting-together process. No chipping, filing, hand tapping, hand studding, hand scraping, drawing over of holes, patching or shimming should be allowed. Whenever a man uses a hammer and chisel, or a file, it means that a blunder has been made somewhere. A mistake has been made in the machine shop, pattern shop or in the drafting room.

Look down the main aisle of the plant. Isn't there a gang of men tramping up and down, some of whom should be working? At the toolroom is not the group of men standing there greater than necessary? Are there not places where two men are working at half speed where one man could do the work of the two?

Watch the foremen. Are they not doing too much clerical work and not enough overseeing? They are hired for overseeing.

If your plant is driven by electricity, take meter readings at your generator every 15 minutes for one day. Note the slow start in the use of power in the morning, also the let down in the late afternoon.

Have all the designing rules that look to economy been followed? Are you eliminating all brass work possible from your designs? Is your design such that you are doing a lot of unnecessary machine work which by closer foundry work could be entirely eliminated? Look at a cast-iron lock. Nearly all the parts in it are just as they came from the foundry. They have practically no machining on them, yet they work perfectly and last indefinitely. See how ingeniously hardware and agricultural machinery are designed in this respect.

Saving in the Foundry

Look at the pile of scrap rejected by the machine shop on account of bad foundry work. This can be reduced. On the most difficult kinds of casting this loss can be kept down as low as 5 per cent., which means that only 5 per cent. of the castings of a difficult nature machined by the machine shop will prove defective on account of bad foundry work.

Is the jar ram molding machine used on all large molds? Is the making of a mold divided into parts, skilled men working on the skilled part of molding only? Are all other parts of mold making done by handy men? Is the squeezer type of machine that makes both cope and drag at one squeeze used on small work? On medium-sized and large work the jarring machine will reduce the molding time 60 to 75 per cent. The squeezer will cut the small molding cost 40 to 50 per cent.

Time your best squeezer molding machine man and your best bench molder on making one mold. Figure at this rate how many molds he should have made. If he is making no unnecessary motions allow him, say 20 per cent. leeway for resting and other incidentals that interfere with the steady run of the work. Is he putting up a good day's work?

Are you running 66 to 75 per cent. scrap, or its equivalent, in off-grade pig iron in the cupola? This is absolutely practicable no matter how fine a grade of castings is being made. Does your iron come glistening white hot from your cupola, thus eliminating dirt and bubbles in your machined faces?

Are you using a mixture for brass with as high a percentage of lead and zinc in it as is permissible on your work? Zinc and lead are very low-priced metals as compared to copper and tin, and the use of them will reduce your brass cost.

Eliminating False Movements

Are the men making false movements? Is their work arranged so that they have eliminated all the unnecessary steps? Instead of giving each man a book of rules and regulations, it might be better to give him a little book teaching him the correct cutting speeds and feeds for different metals, the use of the micrometer index dials on the machine feed screws for duplicating work, and how to eliminate false movements.

Are the total tons output of the plant each year higher than the previous year, without any increase in the hours of labor? There should be an improvement in this respect each year.

Take advantage of all these points, because the profits lie in these small savings throughout the plant. A plant that does not save on all the smallest details will run at a loss, or, at best, just break even. A competitor who just breaks even sets the selling price on your article. He cannot sell any cheaper and continue to exist. Your profit then, if you sell at his price, is what you can save in the little leaks here and there that he neglects.

A 24-Pound Pump of 18 Pieces for $1

A small hand pump is a good illustration of what can be done in economical manufacturing. It is about down to the rock-bottom limit for manufacturing cost. A No. 2 iron cistern pump, with a 1¼-inch suction pipe, weighs 24 pounds, and is sold by the manufacturer to the dealer for $1. This is at the rate of 4 cents per pound.

The pump is made up of 18 pieces, counting the bolts but not the nuts. The pieces are: Handle; spout; 3-inch cylinder; base; forged sucker rod; piston; piston follower; cup leather; leather suction valve; leather suction valve weight; leather suction valve screw; leather suction valve washer; suction pipe reducer; two bolts in handle; one bolt for holding on the spout; two bolts for holding the cylinder to the base. Every piece has machine work on it. The sucker rod has forge work on it. The cylinder is bored and polished its full length.

Just analyze what the $1 that the manufacturer gets for this pump has to cover. We will assume that the profit is 9 cents on the pump, or about 10 per cent. of the cost. Then 91 cents must cover all expenses. Assuming the material and casting cost at 2 cents per pound, which is extremely low, gives a total of 48 cents for the 24 pounds of material. This only leaves 43 cents, about 1¾ cents per pound, to machine, to assemble and to pay office expenses, advertising, overhead expenses and bad debts.

The assumption of 2 cents per pound for the castings, bolts and other material is low for this work. The cheap, heavy handle would be offset by the cored cylinder and lighter pieces. It can be done, though, as foundries sell furnace work at 2¼ cents per pound, stove plate at 2.4 cents per pound and the lighter work at 2¾ cents per pound when pig iron is $17.50 per ton. The foundry's costs are lower than these.

Manufacture for a Bottomless Selling Price

How would you like to sell your product for 4 cents per pound? Go at you manufacturing proposition as though this were the case, and you will make your fortune so fast you will not know where to invest your money. Don't assume that the selling price of a product has reach its bottom; it has not. Firms are making money manufacturing bicycles to-day and selling them to the dealer at $10 apiece. They used to be sold at $50 apiece.

This shows what can be done in price cutting if enough pressure is brought to bear on a product. The output per man in the machine and erecting shop of a certain company increased 84 per cent. and in the foundry 34 per cent. per man in three years. This concern did not take full advantage of its improved position. It was satisfied; satisfied people never do their best. It did not push the sales as it should. It sold more each year, but the sales did not increase as fast as the plant capacity increased from improved methods. The nearest the sales got to the manufacturing capacity in one year was 75 per cent. The number of employees dropped from 230, working overtime, to 165 with no overtime. To-day machines and vises are standing empty, although the business is larger than it ever was before. The interesting thing about this company is that the increase was made purely by management and not by getting in new machinery. In fact nine-tenths of the machine tools were so old that the companies that had made them had gone out of business. It took the firm about 25 years to get rich, whereas it could have done it in 10 or 15 if it had sold to capacity each year.

Spending Money for Equipment

The above is a good example of that type of successful management which lies in working old equipment to the limit, thus increasing the output of the plant without added equipment expense. It is management and not equipment that makes the money. Generally speaking, more money can be made from old machines that are paid for than by borrowing money for new machines to replace the old ones. The heaviest old machines will safely cut 10 cubic inches of cast iron per minute from a casting. This is a good rate.

The old machines have their place. On the lighter classes of work they can keep right up with the modern tools so far as output is concerned. All that is needed is here and there a heavy, high-powered, modern machine to do the hogging work of the plant. New equipment should be of a durable nature, and also of a sort that requires little attention. But no matter what be the economy of a machine or an appliance, if it takes more than usual attention to keep it going, or to keep it in repair it is not practical, and all the advantage of the economy is lost.

A plant full of old tools, which are run up to their limit, will require a big repair gang. Gears will wear out, parts will be constantly breaking, and the machines themselves will get out of line. When the expense of maintenance and repairs exceeds the interest on the cost of a new machine, then it is economy to make replacement, but not before.

When a New Machine is Warranted

Often there is a temptation to persuade ourself that money is being lost on account of a plant being crowded. This may be true, but the interest and depreciation on each square foot of idle floor space probably would amount to more than the extra labor would cost due to being over-crowded. Floor space, including the roof overhead, costs on an average $3.25 per square foot. Depreciation, interest, repairs, light, heat, insurance, etc., at 10 per cent. makes an annual charge of 32.5 cents for each square foot of new floor space, which the new equipment will have to earn before you can come out even. Therefore don't enlarge until you actually have to. Many a firm that has made money in the crowded state has failed after enlarging.

If the interest on the money paid out to make a change in equipment and the depreciation on the tools equals the saving made, nothing has been gained. Suppose a new machine tool is bought to replace an old one. This new tool, costing $5000, will have to average $2.50 profit per day more than the old machine, and continue to earn this through dull times as well as good, because this $5000 could have been put at interest at 5 per cent. with no depreciation in the principal. Depreciation on the tool should be charged at 10 per cent.; interest and depreciation together would be 15 per cent. of $5000, or $2.50 per working day.

Improvements do not always pan out so well as expected, so never buy a machine tool that is to be purely a replacement unless the saving effected by it will pay for the new tool in two years. Follow this rule and you will play safe and still make plenty of improvements.

In making radical and expensive changes, write out a plan in detail and shelve it for a while. New ideas will suggest themselves as time goes on. If possible, go and see the plan or machine where it has been in operation for a year or more.

Buy Equipment in Dull Times

Do all your equipment buying in dull times. They come around often enough. You can get concessions then in price that will make it easier for your equipment to pay for itself. An offer of 5 to 10 per cent. below the regular price will be invariably accepted in hard times.

Choose new machinery carefully. The difference in the durability or in the possible output between two different machines is great. Too much thought cannot be given to this matter of selection. The best is none too good, no matter what the price. New things are changed in some detail after being on the market for a year because weak points develop in the first design. Let the other fellow do the experimenting on the new idea.

The pay-roll is the big expense, and can be cut down by the best machine. A fine machine need turn out but very little more than a poorer one to pay the difference in the first cost, because the life of a machine tool being long, this difference in price is distributed over many years, making the extra charge for each year small.

Put Only Part of the Profits Into Improvements

A certain portion of the profits should be put back into the business in the shape of improvements, but only a portion. Otherwise a manufacturer is liable to live his life putting all his earnings back into his works. All he has at the end is a large works. Neither he nor his family have ever enjoyed any of his earnings. Suppose a firm's yearly profit equals 10 per cent. of the value of the plant, and suppose this firm increases its plant 10 per cent. yearly; then all profits are returned to the plant and none go to dividends. A competitor will forge ahead of a firm that has over-built because the interest and depreciation on idle equipment and idle space absorb profits.

Go through the plants of some of our immense corporations and note the idle cranes and other equipment that are eating up money in interest and depreciation. This is why they have trouble paying dividends. Fortunes are made by the medium-sized manufacturers with seemingly poor and old equipment, whereas some of the big corporations have a hard task at the end of their fiscal year trying to avoid showing a deficit.

Are you putting money into equipment that never pays a cent on the investment? A poor manager will want to spend thousands of dollars for machinery, jigs, tools, changes in buildings, etc., before he can get started. This is the dodge of a man who is not going to make good. The writer knows a manager who is a wonderful and ingenious mechanic who leaves ruin behind whereever he goes. When he takes charge of a plant he spends so much money for tools and equipment that the firm never survives the financial strain. He is always getting the plant ready to manufacture. He never arrives at the manufacturing point. There is always something more to do.

Equipment Which is a Constant Expense

A good illustration of an expensive equipment that makes no saving but is a constant expense is the sand-handling and mixing equipment in some of the newer foundries. We still have to see one that actually decreases the number of foundry laborers. Yet thousands of dollars are tied up in this equipment, and thousands more are spent in power, repair bills and attention just to keep the apparatus running. Each of these equipments takes as much shoveling to get the sand into them and out of them as to cut over the sand by hand. A few husky laborers will cut up a lot of sand in a foundry during the night, with no interest or depreciation charge.

At a foundrymens meeting the principal speaker of the afternoon read a paper on "Modern Foundry Practice." He was the manager of a completely equipped foundry, and said: "To turn out castings economically you must have electric traveling cranes in the foundry and in the yard, overhead trolleys, power boom-derricks, intershop tracks and cars, mold ovens, sand-mixing machinery, sand conveyors to each floor, open gratings in the floor to shake molds out over, lockers for the men's clothing, shower baths, etc." Later, in a discussion on the selling price of castings, this same speaker said: "Something ought to be done to stop the cutting of prices on castings. You do not know what kind of foundries we have up in Chicago. Little low-roofed dark smoky places, down below the street level, where the molders have to work all day by gas light. The way some of those foundries cut prices is a caution. I do not see how they do it and continue to exist. We, with all our fine equipment, lose money when we try to meet their prices." This was an unconscious admission that his company had loaded itself with expensive equipment charges and repair bills that made it hard to meet competitive prices.

The Value of a Night Shift

Do not enlarge until you have to, and then don't do it. It is better to put on a night shift. All the great managers have found that a night shift on machines, with a good hustling foreman, is a gold mine. A night shift increases the output enormously with little increase in the overhead expense. At night three-fourths of the people and material things that make the overhead expense are asleep, and therefore are of no expense to the firm.

The night shift may improve the output 75 per cent. To make this increase by day work would require a large outlay of money, which means additional interest and depreciation charges. Night men require 25 to 30 per cent. higher wages than day men. Set the pay so that men will wish to work on the night shift.

The output per man, at night, should be greater than in the day, because only the modern high-power big output machines are run at night. The men work longer hours, 6 p. m. to 6 a. m., with no shutdown for the midnight meal if this happens to come while the machine is taking a big, long cut. The men are drawing the highest wages in the plant and are not likely to shirk.

It does not pay to run a night shift more than five nights a week. A night shift cannot be used in assembling for quality will suffer. In the foundry it pays to run pouring and shake-out gangs at night.

The output of a plant is controlled by the machine department. All other departments are flexible. Enlarge this gateway and you have increased the whole plant capacity. A plant rushed to the breakdown point, the foremen crowded with work, and each workman with ten jobs waiting to get into his machine is in the most desirable state.

Cutting labor cost is not of the first importance if it means cutting wages. A high output per man, even with high wages, will reduce the cost per piece and enlarge the capacity of the shop. The good manager will seek to attain this end. For that is the pathway to profits.

ARTICLE XIV

INCREASING SHOP PRODUCTION

Illustrations of Efforts to Bring Machines to Maximum Capacity with the Resultant Effect on Overhead Charges and Profits

THE actual output of the average cotton mill is from 80 to 90 per cent. of its theoretical capacity; that of the machine shop is seldom over 30 per cent. With proper selection and training of the men, and with good management, a machine shop can attain the same efficiency as the cotton mill, or nearly three times its present capacity. To accomplish this, however, those in authority must plan the work, the fixtures and shape of cutting tools instead of leaving all these to the workman. Thirty per cent. sounds like a low figure. The following incident shows how nearly right it is.

Certain machines in a shop had a month's work ahead. One machine, a cylinder boring lathe, was using a feed of 1/32 inch per revolution. The foreman ordered it increased to 1/16 inch. The workman, as soon as the foreman's back was turned, dropped the feed again to 1/32 inch. Continued pressure on the man brought the feed and speed up to the apparent limit of capacity of the machine.

The superintendent offered the machinist, who already was receiving high wages, an increase of 1½ cents per hour. At the same time he told the man that he was going to watch him and ascertain if the shop was receiving an increase in output over the present rate equivalent to, or greater than, the increase in pay. The workman promptly rose to the occasion, and the noise made by the machine on the roughing cut could be heard all over the shop. But little time was lost in changing cutters and only a few minutes were needed to change cylinders. For finishing the ends of the cylinders two-bar facing attachments, one at each end, were used on the boring bar. The workman put a belt tightener on the cone belt to enable the machine to pull a heavy cut. The net result was nearly double the previous output which had been the apparent limit.

Another incident will bear out the statement made in the first paragraph. A number of 12-inch steam cylinders, 15½ inches long with a 3/16-inch finish to be removed on each side on the roughing cut, and with 1/32 inch to be removed on the finishing cut, were to be machined in a Draper cylinder lathe about ten years old. The foreman decided to see just how quickly the job could be done. The machine was speeded to 14 r.p.m. A roughing cut of ⅛-inch feed per revolution was taken with a feed of ¼ inch for finishing. The machinist objected, saying that the belt would slip under such a cut and that the cylinder would spring so much under the roughing cut that the finishing cut would never true it up. Neither thing happened. The belt did not slip and the finishing cut trued the bore perfectly.

What Working a Machine to Capacity Accomplished

This incident shows that the average workman does not know the capacity of his machine. The machinist in this case really believed what he said. The roughing cut on this job required 11 minutes and the finishing cut 6 minutes. Fifteen minutes were necessary for removing the finished casting from the jig and putting in a rough one. One minute was spent in tool changing; a total of 33 minutes for the job. The best previous time for this operation was 95 minutes.

Let us analyze what this means to the firm. If the time on every operation in the plant could be cut in the same proportion, the time and labor cost could be cut 66 per cent. The firm, in the past, we will say, was making a 10 per cent. profit. Assume that the average finished machine cost was as follows:

Material in machine	$ 82.00
Labor	20.00
Overhead and selling expense	51.00
Total cost on machines	$153.00
Selling price	$168.30
The cost	153.00
Profit, 10 per cent	$15.30

Now, cutting the above labor cost to 34 per cent. of what it had been, would change the cost to this:

Material, as before	$ 82.00
Labor—34 per cent. of former cost	6.80
Overhead and selling expense—75 per cent. of former cost	38 25
	$127.05
Selling price, as before	$168.30
Cost	127.05
Profit	$41.25

Instead of a profit of 10 per cent. on each machine shipped, the profit would be 32 per cent. More than that, the increased output per man and per machine would mean an increased yearly plant capacity of 294 per cent., with no increase in the equipment.

Suppose the sales were $110,000 per year, with a cost of $100,000, or a profit of 10 per cent. Now, with the increase in the output of 294 per cent., the sales would be $323,400. At the new rate the cost would be $245,000, leaving a profit of $78,400, or 32 per cent.

As an actual fact, the showing would not be as great as this, for in order to triple the sales the selling price would be cut. This would reduce the profits. Another point that would reduce the profits would be an increase in the plant pay-roll. Each productive workman, on account of the added strain on him from increased production would have to be paid 20 to 30 per cent. more than in the past. This would affect only that portion of the force that were actually doing the work on the product. It would probably increase the pay-roll 10 per cent.

Suppose these reductions brought the year's profits down to $50,000 or $60,000. Comparing this with their previous profit of $10,000, it shows that a firm can make a comfortable fortune in 10 or 20 years if it will go to the limit in manufacturing and selling.

Changes Possible in Overhead Charges

The following changes in overhead expense would be made if the output were tripled with no increase in the equipment or the number of men who do actual work on the product. Before tripling the assumption is that each $100 of overhead expenses is divided as shown in the first column.

	Before	After
Office salaries	$ 16.50	$ 24.00
Traveling men's salaries and expenses	16.00	38.10
Advertising	5.50	14.45
Office supplies and catalogues	3.00	9.00
Factory heads' salaries	7.80	9.80
Pattern expense	8.30	16.60
Petty cash, freight, drayage, etc.	9.50	27.00
Tool supplies, shop castings, etc.	3.20	9.00
Work on machine tools and jigs	9.50	26.00
Roustabouts	6.20	17.50
Studs, bolts, paint and like supplies	8.30	24.00
Engine room labor	3.00	3.25
Coal bill, belts, etc.	3.20	6.30
	$100 00	$225 00

Where $100 had been spent as overhead expense manufacturing a machine, $225 would be spent on three of these machines as overhead expense, after the plant output had been tripled, with no increase in the equipment. The overhead expense on one would be $75. Thus the overhead expense on any one machine, built after the increased output, would be only three-fourths of what it had been before.

The only increase in the office force would be in the circular letter-writing department, and possibly an extra person to answer the added correspondence. The increase in the cost-keeping and bookkeeping departments would be very little. A couple of low-priced girls could take care of it.

The traveling expenses would not be tripled, as certain economies could be practiced when selling on a large scale. The low selling price would be a great stimulus to the sales department. The advertising field could be fairly well covered without tripling the advertising. The office soliciting and catalogue expense would be tripled, as letter soliciting would be pushed to the limit. The number of factory heads would not increase. Their salaries would be a little higher.

Pattern expense would not increase in proportion to the increased output, as the increase would probably come on the more standard lines of work, which would require no pattern work. The tool supplies, shop castings, etc., would not triple, nor would the tool work be tripled. Roustabouts would not increase three times.

Studs, bolts, nuts, paint and like supplies would not triple in cost, as on account of buying in larger quantities lower prices would be paid. Some supplies that previously had been bought would be manufactured.

Engine room labor would increase little, if any. The coal bill would increase very little. There would be no more power used to overcome the dead-load friction (shafting, countershafts, loose pulleys, etc.) than before. There would be no increase in the coal for heating.

Ideal Conditions in the Machine Shop

There are certain ideal conditions that every machine shop should strive to attain. These are. (1) all machine tools should cut to the limit of the power of the tool steel; (2) the chucking fixtures should be so designed that the chucking time on machines is reduced to a minimum. The West Albany Shops of the New York Central & Hudson River Railroad furnished a fine example of the attainment of these ideal conditions. The information was obtained from the Niles-Bement-Pond Co.

The job was the turning of 36-inch Krupp & Paige wheels in a Pond lathe. During a continuous run from 7:00 a. m. to 5:35 p. m., with one hour out at noon, 33 pairs of wheels were turned. The details of the performance are shown in the table below:

Average time putting wheels into lathe	2 min. 28 sec.
Average time roughing	9 min. 23 sec.
Average time finishing	5 min. 17 sec.
Time required for removal	1 min.
Total average time on one pair of wheels	17 min. 58 sec.
Average depth of cut	3/16 in.
Average cutting speed	14.4 ft. per min
Cubic inches of steel removed per minute average	13.1
Minimum time putting wheels in lathe	1 min.
Maximum time putting wheels in lathe	4 min.
Minimum roughing time	7 min.
Maximum roughing time	12 min.
Minimum finishing time	3 min.
Maximum finishing time	7 min.
Minimum time on one pair of wheels	14 min.
Maximum time on one pair of wheels	21 min.

These ideal records were obtained after fitting the lathe with a pneumatic tool-clamping holder, power movement of the heads on the lathe bed, pneumatic clamping of the heads to the beds, by having the segment in central driving gear open, close and lock automatically when wheels were rolled in and out—no attention being required by the operator. These devices reduced the idle time of the lathe to 3 minutes 28 seconds, or 11 per cent. of the total time.

By having a machine powerful enough to cut to the limit of the tool steel, by removing the finished work promptly from the vicinity of the machine, and by having raw material at hand for the lathe so that no time was lost waiting for it, an output of 30 pairs of tires per day of 10 hours was secured day after day. The machine in question was driven by a 40-hp. motor, which is equal to one-half the power used at the tool points in the average machine shop employing 250 men.

Manufacturing Losses that Never Show

A firm must get its manufacturing costs low enough to be able to sell a little under the market and still make profit. It must be a good, safe profit, as there are manufacturing losses that never show on the cost cards. The actual cost of a finished machine is higher than that given by the cost cards. This fact will be noticed at the end of the year. Some of these losses are: The scrapping of rough or finished parts on account of a change in the design; change in the market that makes a certain class of machinery, or parts of machines, obsolete. These parts may be retained, but the loss is there just the same. Money has been spent for something that will never bring a return.

To select the best method of machining a piece analyze and time the different steps taken. For instance, the number of seconds required to chuck the piece; the number of seconds to put in the roughing cutter; the number of seconds required for the roughing cut; the number of seconds necessary to change cutters; the number of seconds required for the finishing cut; the number of seconds required for taking out finishing cutter; the time necessary for taking the piece out of the machine, etc. Having this information concerning several methods, one can easily decide what is the best method of doing the work or devise a new method better than any of the old ones. The method having the least idle time will be the best.

Avoid as much as possible the use of special machine tools. Use the method that will allow a small number of jigs and fixtures to cover all sizes of pieces. Select that method which will enable the operator to be setting up one piece while the machine is cutting on another piece.

Determining on Best Method of Operation

A good method of investigation is to take the most popular size of machine that you build. Get out the rough stock complete for two machines. Start machining these a piece at a time. Note in writing every operation and every step in each operation down to the minutest detail together with the time required for each. Do the same in the assembling and the erecting. Surprisingly bad methods will be discovered the correction of which will cut the cost and increase the plant capacity.

The statistical information thus acquired will always be useful in checking the time and cost sent in by the men on similar operations or in deciding whether or not a change in design will decrease the cost. It will tell whether or not the making of a jig or fixture will pay, or whether a new machine tool will improve the output, and how much it will improve it. The spending of money for improvements then will be a safe proposition. It will be known that the money will come back. To know is better than to guess, or to take the guessing of others.

These notes may be used as a basis for putting the operating time, speeds and feeds, etc., upon the drawings in the forms shown in Table I.

Table I—Information Tabulated in Studying a Given Machine's Production

Setting number	Operation number	Operation	Surface machined	Depth of cut, in.	Feed per revolution	Rev. per minute	Minutes each operation	Minutes actually required
1	1	Chuck.					½	½
	2	Rough turn } simultaneous cuts	A	⅛	0.111	8	3	3
		Rough face	B	⅛	0.111	8	1½	
		Rough face	C	⅛	0.111	8	1½	
	3	Finish turn } simultaneous cuts. .	A	0.005	0.333	8	1½	1½
		Finish face	B	0.005	0.333	8	¾	
		Finish face	C	0.005	0.333	8	¾	
		Rough bore } simultaneous cuts	D	5/16	0.111	45	2	2
		Round edges				8	¾	
	5	Truing cut.	D	1/64	0.056	53	1	1
	6	Ream.	D	0.005	0.216	30.7	¼	¼
	7	Remove.					½	½
2	1	Chuck.					½	½
	2	Rough face } simultaneous cuts. .	E	⅛	0.111	8	1½	1½
		Rough face	F	⅛	0.111	8	1	
	3	Finish face.	E	1/64	0.333	8	½	½
	4	. . . .	F	1/64	0.333	8	½	½
	5	Round edges.					¾	¾
	6	Remove.					½	½
		Total time (individual operations)	. . . .	. . . .	. . . .	. . . .	19¼	
		Actual time.	. . . .	. . . .	. . . .	. . . .		13

These instruction sheets should be pasted on the drawings. They give, in the minutest detail, not only each step in each operation, but the jigs, tools and fixtures, speeds, feeds and depths of cut, the time required for and the sequence of operations. The jigs and tools will be numbered and referred to by number.

Many jigs are made at great expense and never used a second time because they are forgotten. This is especially true where the jig is made to save time on some minor operation. The system of having a complete instruction card on each drawing will eliminate this wasteful condition. Pasting the instructions on the drawing will allow changes to be made in this sheet without making a new drawing each time. This system will allow the chief of the producing end to be a hustler rather than a fine mechanic or engineer.

ARTICLE XV

CUTTING THE COST OF POWER

Economies Effected by the Judicious Selection and Use of Electric Motors or Other Equipment—Hints on the Prevention of Smoke

DRIVE all small tools requiring less than 5 hp. by belts from a line shaft. Drive all tools requiring 5 hp. or more by individual motors.

Never use the group system of electric drive except in an isolated department where the group motor can be shut down half the time. For instance, the pattern shop, where the men are doing handwork most of the time and all the machines are idle a great part of the time is a good place for the group drive.

More power is wasted and lost in driving a machine by electricity than by belts and shafting direct from the engine. This is true during the time the machine is running. When it is shut down the reverse is true, as then a great amount of power is being wasted by the shaft-drive system. To get any benefit from electric drive, the machines must be shut down a portion of the time. An individual motor-driven machine that runs continuously will have a greater power loss than that same machine driven direct from the engine through belts and shafting.

Power Loss from Engine to Tool Point

Fifteen per cent. of the energy is lost in a generator running under half load. A generator that is receiving 100 hp. from the engine will deliver but 85 hp. to the switchboard. Five per cent. of the energy is lost in transmitting the electricity through the lead wires in a plant, so that of the 85 hp. at the switchboard, but 80¾ hp. is available at the motors.

If the motors are running on one-third load, which is generally the case, the power loss, changing from electrical power to mechanical power, in the motors will be about 17½ per cent. The 80¾ hp. at the motors will drop to, roughly, 66½ hp. mechanical power at the motor pulley or gear, a net loss of 33 1/3 per cent. from engine to motor armature.

The loss in each pair of gears and journals will average 7 to 10 per cent. The friction in well-cut gears will absorb about 3 or 4 per cent. Adding the friction in the bearings or journals the friction in the machine will be about 7 to 8 per cent. on the average. As the gears wear the power loss increases. Thus 66 2/3 hp. at the driving pulley of old machines, will, when transmitted through five pairs of gears in each machine tool, shrink to about 33 hp. at the tool points. When driving electrically therefore over two-thirds of the power is lost before it gets to the tool points. The only thing that saves the electric drive is the fact that power waste completely stops when the motors are shut down.

The loss when driving tools by shafting and belts will run about 50 per cent. in the average case. About 5 per cent. of the power can be saved by changing the oil in all the lineshaft bearings four time a year, instead of once a year. Roller bearings will cut the friction loss in half. It is practical to use them on new installations only It is best to try a couple of roller bearings in a very hard place for a year or two to make sure of getting a durable style. A bearing that will not wear well must be avoided, as a shutdown due to trouble with the line shafting is very expensive.

Advantages of the Three-Phase Motor

Use three-phase alternating-current induction motors on all equipment except the cranes. Use direct-current motors on the cranes.

The alternating-current motor is cheaper than the direct-current motor. It will stand a bigger overload. The cost of repairing, when burned out from an overload, is less than the cost of repairs on a direct-current motor. The greatest point of all is that the alternating-current motor requires no more attention or expense for up-keep than the old-fashioned grindstone. Two ring-oiling bearings are the only points of wear. The electric current goes into the stationary part of the machine only. The revolving part of the machine has no wire on it at all, so that the troublesome brushes and commutator are entirely eliminated. A plant fully equipped with alternating-current induction motors will have no trouble, whereas one equipped with direct-current motors, especially where there is iron dust in the air, will have to keep a man continually busy repairing short circuits on the commutators and fixing up the brushes.

Alternating-current motors are not adaptable to variable speeds, and for this reason they are not satisfactory for cranes or for doing work where they have to be run slowly at times; but for all other work, they are ideal.

Motor Costs and Efficiencies

A 5-hp. alternating-current motor costing $64 will safely do the work of a 7½-hp. direct-current motor costing $156. A 7½-hp. alternating-current motor costing $121.50 will do the work of a 10-hp. direct-current motor costing $166. The reason for this is because there is nothing on an alternating-current motor to spark and burn. Alternating-current motors of 5 hp. and smaller require no starting boxes. For this reason they are cheaper per horsepower than larger motors.

The author knows of a firm which buys nothing but 5-hp. motors, or smaller. They put two motors on one machine if they find that one motor fails to pull. Two 5-hp. motors cost $128. One 10-hp. motor costs $166.

A destructive overload on an alternating-current motor makes itself evident in the shape of heat in the motor. The solder starts to fly out into the field winding. A 75-hp. motor, if burned out, will cost only $40 to repair, so that it is best to risk putting in motors that are a little small for the work to save first cost, and thus get a better power efficiency. The overloaded motor uses electricity economically. The motor running light is extravagant in the use of electricity.

The makers will guarantee motors to stand a 25 per cent. overload for 2 hours. Motors will actually stand an overload of 25 per cent. for 4 hours; 50 per cent. for 1 hour, and 75 per cent. for 10 minutes. This would mean that a 7½-hp. motor can deliver 9.4 hp. for 4 hours; 11.3 hp. for 1 hour and 13.1 hp. for 10 minutes.

The efficiency of a motor is the percentage of the electric energy delivered to the motor that is turned into mechanical energy.

The efficiency of a 7½-hp. motor on different loads is:

1/10 load, or 3/4 hp.	about 60 per cent.
3/4 load, or 2½ hp.	about 82½ per cent.
3/4 load, or 5½ hp.	about 88 per cent.
Full load, or 7½ hp.	about 88 per cent.
1/4 overload, or 9.4 hp.	about 87 per cent.
1/2 overload, or 11.3 hp.	about 87 per cent.

Larger motors have slightly better efficiency; thus the efficiency of a 75-hp. motor is:

½ load, or 37½ hp.	89 per cent.
¾ load, or 57 hp.	90 per cent.
Full load, or 75 hp.	90 per cent.
¼ overload, 94 hp.	89 per cent.

The efficiency of direct-current motors is about the same as that of alternating-current motors. The same is true of generators. An alternating-current generator will stand a 50 per cent. overload for 2 hours; 75 per cent. overload for 1 hour and a 100 per cent. overload for 1 second.

The following efficiencies can be obtained with generators:

	Efficiency, per cent.	
Percentage of full load	30-kw. generator	50-kw. generator
100	87.4	90 5
75	86 0	89 0
50	85 5	85 0
25	81.0	83 0
10	50 0	50.0

When figuring on the size of an alternating-current generator to be used for driving motors, a margin must be allowed for power factor. A larger generator has to be installed than would be necessary if there was no such thing as power factor. Power factor does not increase the load on the engine. Its effect is purely local in the generator.

A plant with a 100-hp. compound condensing engine run in a somewhat slipshod way, with the engine in rather bad condition, the boiler setting leaking air more or less, the feed water not heated with steam from the auxiliaries, can make power for less than 2 cents per kilowatt hour.

Smoke Prevention

Smoke can be prevented or reduced by the observance of a few simple rules.

1. Fire five shovels of coal on one side of the furnace, covering the fire evenly and keeping the fire level. Five minutes later five shovels of coal should be fired on the other side in the same way. Keep this up as long as the demand for steam is heavy. As the demand for power

decreases, reduce the number of shovels at each of the 5-minute periods, but *do not lengthen the space of time* between firings, until the call for power is so light that two shovelfuls are enough every 5 minutes. If this rate of firing still gives too much steam, lengthen the time between firings. After each firing leave the fire door open about 2 inches for 1 minute, or until the smoke-producing gases have left the coal. Fasten a door check upon the boiler front in such a way that it will swing around slowly and close the fire door automatically after the heavy smoke producing gases have passed off the coal. This apparatus makes a very efficient smoke preventor.

2. Shake the shaker-grate once an hour. Do not overdo this, otherwise the grate bars will be burnt, and unburned coal wasted through the grate.

3. Clean the boiler flues once a day, either before starting in the morning or during the noon hour.

The foregoing rules of firing will reduce smoke to practically nothing and will keep the fires clean, which is economical of coal.

It is a good idea to rig up an alarm clock to ring electrically every 5 minutes to notify the fireman of the exact firing time. Solder a long finger on the minute-hand setting the knob at the back of the clock. Arrange this finger to make the electric contact at 5-minute intervals to ring a bell.

ARTICLE XVI

LOWERING MACHINE WORK EXPENSE

Use of Automatic Machines and Tool Holders—Establishing Fitting Allowances—Things Which the Machine Designer Should Consider

GENERALLY speaking, there are few jobs that an automatic or semi-automatic machine will turn out faster than a standard type of machine, provided the standard type is as powerful as the semi-automatic in drive and has quick changes of speed and feed. In other words, the greatest advantage of the special machine is its power and quick changes of feed and speed.

Automatic Versus Standard Machines

A powerful, standard machine, provided with quick changes of speed and feed, will generally turn out work cheaper than an automatic or semi-automatic, because:

1. The average machine job, having few operations on it (it should be redesigned if this is not so), rarely brings the special features of the semi-automatic into play.

2. In general, the number of pieces to be run through are too few to pay to set the stops, etc.

3. The machines seldom are properly equipped, because each new job generally takes a complete outfit of special tools.

4. On account of these machines being more complicated than the average standard tools, often some one part is out of order, not badly enough to pay for overhauling, but enough to interfere with the running.

5. The price of such a machine is generally prohibitive. It is safe to say that in three cases out of four, expensive machines are run the same as common machines, thus getting no benefit from their special features.

A powerful engine lathe equipped with a four-sided turret tool post and plain cross feed stops will turn out duplicate work very rapidly. Add a home-made bar holding chuck and the machine is practically equal to a very expensive bar stock machine. Longitudinal stops are not very necessary as length is easily gauged while the lathe is cutting. A clamp can be put on bed for one stop and the tools located at different points in the turret to cut shoulders at various points.

Fitting Allowances

Make a table giving the allowable looseness of the different parts on the machine you manufacture. Some parts do not require a close fit. On these a saving will be made by not working too close. Accuracy, as a rule, cuts down speed and increases costs. Put this matter down in black and white, else the workmen will have no way of knowing what work is particular and what is not.

Wherever possible have each jig so made that it tests th accuracy of previous operations. Make the jig so that it will not fit the piece or so that the piece will not go into the jig if the previous operations have been done inaccurately. This method discovers faults in workmanship immediately, without the expense of an inspector.

Number each jig and put these numbers on the drawings. This will insure the jigs being used.

Three Machine Shop Suggestions

A boring bar guided close to the work will stop all chatter, and will allow an increase in output of about 30 per cent. over boring with a bar not so supported. The casting can be much out of true in the rough and not affect the speed of production. There will be no necessity for cutting down the speed or feed in taking the finishing cut. In fact, the feed on the finishing cut can be doubled and still make a smooth job. The boring can be done in two cuts instead of three. This is not possible when the bar is not so guided.

On duplicate work, where the tool need not be disturbed, put a clamp on the lathe to act as a stop for the cross slide. Accurate work can be turned out very rapidly by this method, as it eliminates calipering.

The planer type milling machine, with a number of cutter heads, reduces the milling cost on work having two or more faces to be machined, provided there is enough work to keep the machine continually busy.

Tool Holders

Use tool holders on all small and old machine tools. Probably 95 per cent. of the work done in a machine shop is light enough to be handled with tool holders instead of solid tools. They are great savers of tool steel and save even more in wages which must be paid out for forging and grinding solid tools. They keep the man at the machine.

A 3/4-inch bar of square steel on a long continuous cut on rather hard cast iron, at 70 feet per minute, a 1/9-inch feed, and a 1/8-inch depth of cut, will remove 11 cubic inches per minute. A stream of chips, literally red hot chips, will come off the tool point.

On soft cast iron, a good grade of 3/4-inch square high-speed tool steel will run 110 feet per minute, 1/9-inch feed, and a depth of cut of 11/64 inch, removing at this rate 25 cubic inches per minute.

Thus there are few places where the full size forged tools are needed.

I know of a firm that uses $\frac{3}{4}$-inch steel with no tool holders. They place a piece under the bar of tool steel in the tool post of the lathe to bring the height of the steel up to the correct point. This works very satisfactorily.

Any firm that uses a good grade of high-speed steel in this way will have a smaller tool steel bill than in the days before high-speed tool steel. The reason for this is that high-speed steel is ground very little as compared to the carbon steels. This freedom from grinding is just as true of high-speed steel drills, especially the twisted types, as it is of cutters.

General Machine Specifications to Be Laid Down

The product of the designer must be an article that is readily salable as well as easily made. Certain general specifications will have to be kept in mind, which make a strong, smooth-running easily operated machine. The more important points follow:

Have quick speed changes. Have quick feed changes. Speed steps should not be greater than 50 per cent. from one step to the next that is, 2 to 3. Have all operating handles within easy reach of the operator.

Have power pass through as few gears as possible from motor to work. No gear should run higher than 1000 feet per minute tooth speed. Above this speed use belts or the silent chain. All gears must be protected. Worm gear drives should be avoided, as they wear rapidly and require excessive power. Machine tool makers, on some of their heavy machines, are cutting the gear teeth on an angle. This prevents chatter and the marking of the work by the gear teeth.

Make all sliding surfaces, such as ways, etc., of ample area to take care of thrust pressure and wear. Have lubrication well taken care of. On all high-speed bearings use ring oilers.

Equip all feed screws, etc., with index dials graduated to 0.001 inch so that accurate movements can be made without calipering or measuring the work. In addition, duplicate cuts on different pieces can be taken without waste of time.

Make all bolt slots so that the depth of the narrow part is at least 30 per cent. greater than the width of the slot, to prevent the breaking of slots.

It is becoming the rule on lathes to design the spindle one-quarter the diameter of the lathe swing for machines that take enormous cuts. The tailstock barrel is made one-sixth the swing of the lathe. The tailstock is locked down with plenty of heavy bolts.

In choosing the motor for a machine, a rough rule is to allow 1/2 hp. to remove 1 cubic inch per minute of cast iron, and 1 hp. to remove the same amount of steel. This estimate may be 50 per cent. in error either way, depending on the shape of the cut, the hardness of the metal and the loss of power in the gearing, bearings, etc.

The power lost in gearing is greater than is usually realized and for this reason slow speed motors are desirable. To remove 126 cubic inches of steel per minute in an engine lathe at a cutting speed of 28 feet per minute 1 1/2-inch depth of cut, 1/4-inch feed per revolution, using a single tool, 80 hp. is required. To reduce a bar 2 1/4 inches in diameter to 1 inch, with 11 3/8-inch feed per minute, or at the rate of 35 cubic inches per minute, in an engine lathe, requires 20 hp. About 92 hp. is required to remove 400 cubic inches of steel per minute on a milling machine.

Tool Pressure

The tool pressure on a lathe is 75 tons when removing 11 cubic inches per minute with a cut 3/8 inch deep, and 0.05-inch feed per revolution at 185 feet per minute cutting speed. This is on mild steel having a tensile strength of 56,000 pounds and 44 per cent. elongation. The end thrust when drilling with a 3-inch drill, at 100 r.p.m. and 0.03-inch feed per revolution or 3 inches deep per minute, is about 5000 pounds. Such work would require a 50-hp. motor on the machine.

ARTICLE XVII

CAPACITIES OF MACHINE TOOLS

Ways in Which the Superintendent Can Learn the Most About Feeds and Speeds—Some Unpublished Tests—The Shape of Tools

A CUTTING speed of 100 to 120 feet per minute is about correct for outside turning of soft steel, while 75 to 80 feet per minute is about right for soft cast iron. For inside or boring work, 40 to 60 feet is the proper figure, and 500 feet per minute is the speed for brass turning. Recently a satisfactory high-speed steel for brass has been made. The tool steel maker is improving tool steel so rapidly that these speeds probably will be increased 15 per cent. in the next five years.

Tool Steel Tests

For testing tool steels a casting made especially for this purpose should be kept on hand. It will be a massive disk of cast iron, say 18 inches in diameter and 8 inches thick, with bolting-down ears cast on it, and a tapped hole through the center for a lifting eye-bolt. The analysis of the iron in the test casting used by the writer is as follows:

Silicon	1.480 per cent.
Manganese	1.070 per cent.
Graphitic carbon	2.560 per cent.
Combined carbon	0.910 per cent.
Phosphorous	0.328 per cent.
Sulphur	0.069 per cent.

This, however, is not the mixture used in the regular castings, the silicon being lower, the manganese higher and the sulphur lower. This casting was made especially to test tools on. Any mixture would do as the test is a competitive one of each tool against every other tool.

Test the tool steels by starting a facing cut at the center of the disk and face out toward the periphery. Adopt a certain standard speed, feed and depth of cut. The different kinds of tool steels will break down at different diameters. The steel that stands up on the cut of greatest diameter is the one to adopt. This method of testing gives a true result. The test piece is the same for all steels. It is handy to set up and take out of the machine. It is the only good way to settle the tool steel question.

By means of this test piece and other observations described later, the writer had cutting speed tables for Mushet self-hardening steel pretty well made up before F. W. Taylor published the results of his experiments.

As a result of his tests the author patented a cutting tool using self-hardening steel that would cut nearly as fast as the present high-speed steel. A piece of Mushet steel was cast in a heavy copper holder so that only the bare cutting point or edge of the steel showed outside of the copper. The copper carried away the heat as fast as it was generated at the point and thus allowed an increase in cutting speed. About a year after this patent issued Taylor and White's articles on their tool steel experiments at the Bethlehem Steel Works were published, and this resulted in our dropping the tool holder and adopting high-speed steel when it came onto the market.

The Best Shape of Tool

The best shape tool is one ground to a round nose at the end. The top of tool is the shape of the end of a man's thumb. The top surface should slant some so as to give a lifting effect on the chips, not a scraping effect.

The clearance between the end of the tool and the work should be very small.

We now use high-speed cutting steel altogether in our work, except on brass. On this we use self-hardening steel. High-speed steel fails to hold a keen edge and cannot be used on brass, as a dull tool on brass will burnish the work.

Things the Superintendent Should Know

The superintendent must become not only a hard-working student on the subject of speeds and feeds and cubic inches per minute; he must become an authority on it. He must know the exact capacity of each machine tool in his plant. He cannot carry all of this knowledge in his head, so he should make out a book ruled as here

Feet Per Min.	Total Cubic Inch Per Min.	Cubic Inch Per Min. Each Cutter	Total Feed	Feed For Each Cutter	Depth	Roughing or Finishing Cut	Diam. of Work	Name and Shape of Tool Steel	Remarks

shown and having the same headings across the top of the pages.

Accumulating Machine Data

Each machine in the plant should have a separate page or a number of separate pages in this book to represent it. The pages should be numbered with the same numbers that have been given the machines. This book should be of the loose-leaf type, as on some of the important machines there will be three or four pages of these statistical notes. For instance, the cylinder boring lathe page will read as shown in Table II.

The space under "Remarks" will gradually fill with such notes as: Limit of power of the machine; hard cast iron; tool O. K.; cut ten minutes; made machine quiver; broke driver; very heavy cut; belt broke after nine min-

Table II.—Cylinder Boring Lathe Data

Feet per Min.	Total Cu. In. of Each	EACH TOOL			Rough or Finish	Diam., In.	Name and Shape of Tools	No. of Tools	REMARKS 4x¼ in. Belt
		Cu. In. each Cutter	Feed In.	Depth In.					
22	6.14	6.14	⅛	$\frac{3}{16}$	Rough	$5\frac{11}{16}$ raised to $6\frac{3}{32}$	....	1	
19	....	0.44	$\frac{1}{16}$	$\frac{1}{32}$	Finish	14⅞	Novo	......	
17½	....	1.5	1–48	$\frac{1}{32}$	....	....	....	......	
35	29½	14.75	⅛	$\frac{3}{32}$	Rough	16	Rex AA	2	¼ in. total feed 8 r.p.m. 9½ min. on cut. Beyond belt limit. Held belt with a stick. Stalled five times. Limit of tool steel. Chips size of corn and cut.
35	6.58	3.28	⅛	$\frac{1}{16}$	Second	16	Burgess No. 5	2	¼ in. total feed 9½ min. on cut.
30	5.6	2.82	$\frac{1}{32}$	¼	Rough	$8\frac{3}{16}$	Burgess No. 5	2	$\frac{1}{16}$ in. total feed 14 r.p.m.
39	7.4	3.6	$\frac{1}{16}$	⅛	Rough	18	Burgess No. 5	2	⅛ in. total feed 8 r.p.m. HARD. Cylinder. Limit of stool steel.
30	8.5	4.25	$\frac{1}{16}$	$\frac{3}{16}$	Rough	8	Burgess No. 5	2	⅛ in. feed 14½ r.p.m.
33	6	3	$\frac{1}{16}$	⅛	Rough	16	Burgess No. 5	2	⅛ in. feed.
35	24	6.15	$\frac{3}{16}$	$\frac{5}{32}$	Rough	18	....	4	Double rig ⅜ in. total feed. PACE.
26½	15	....	$\frac{1}{64}$	⅜	Rough	12	....	4	Double rig but no chucking jig. Two separate cylinders. $\frac{1}{32}$ total feed.
	Belt would just pull chip. Chips $\frac{3}{32}$ in. thick, ½ in. wide on ½ in. circle.								
45	12⅔	3	$\frac{1}{32}$	$\frac{3}{16}$	Rough	12	....	4	Double rig. Chucking jig.
	$\frac{1}{16}$ total feed. Belt O. K. 15 min. chucking 21 min. each cut. Has to loosen up jig for fin cut. Jig could be designed stiffer.								
62	26	13	$\frac{1}{32}$	$\frac{3}{16}$	Rough	16	....	2	15 r.p.m. $\frac{1}{16}$ in. total feed. Finished next speed slower ¼ feed. Fine finish.
	15⅞ in. rough 15W a8ter cut. One bar. Chips ⅞ in. wide, $\frac{1}{16}$ in. thick. Roughing cut jars floor at foreman's desk.								

Table II—Cylinder Boring Lathe Data—(Continued)

Feet per Min.	Total Cu. In. of Each	EACH TOOL Cu. In. each Cutter	Feed In.	Depth In.	Rough or Finish	Diam., In.	Name and Shape of Tools	No. of Tools	REMARKS 4x¼ in. Belt
46	7	....	..	13/64	Rough	12	....	......	14½ r.p.m. ⅛x1/16x7/16 in., size of each viiual chief grain.
	11 17/32 in.	rough,	11 13/16 af	ter cut	1/16 total	feed.			
57	5	2½	1/32	⅛	Second	14¼	....	2	15 r.p.m. Belt loose and slipping slightly. 1/16 total f ed.
46	4.29	2.15	1/32	⅛	Rough	12	Burgess No.5	2	1/16 in. total feed 1–256 area of cut. SLOW.
45	4.20	2.1	1/32	⅛	Rough	12	¾x¾ Novo	2	1/16 in. total feed 6 in. x 8 ft. bar. 12 in. cylinder.
37½	14	3.51	1/16	⅛	Rough C. I. bush	Two 9½ bushes	½ Sq. Burgess	4 2 in a bar	⅛ in. total feed. Double rig. 15 r.p.m. Just could bear hd on ush. Hot. 7 cu. in. each side. Belt 1,247 ft. per min.
PACE	FOR M	ACHIN	E.						
36	6.75	1.69	1/32	⅛	Rough	10	....	4 2 in each	1/16 total feed.- 14 r.p.m. Double rig. 2 cutters each bar.
33½	6.25	6.25	1/16	¼	Rough	16	Self hard	1	NOTE 1 Single erd scrap ctr.
30	8.4	4.2	⅛	3/32	Rough	14	Burgess No. 5	2	¼ in. total feed. p.m. 2 cuts only 8 min. dura of cut.
30	5.6	2.8	¼	1/32	......	......	....	2	½ in. feed total. .p.m. 4 min. ut.
PACE	Finish								
30	16.8	8.4	1/16	5/32	Rough	7¼	....	2	⅛ in. total feed. 15 r.p.m. Held belt with a stick. Chips ½x½x3/32 in. Very soft iron.
PACE	Beyond	belt pull.							
33	12	6.1	1/16	⅛	Rough	16	Burgess No. 5	2	⅛ in. total feed. 8 r.p.m.
33	6.1	6.1	1/16	⅛	Rough	16	Burgess No. 4	1	8 r.p.m. Note single end ι ter.
30	5.62	2.8	1/32	¼	Rough	14	Burgess No. 5	2	1/16 in. total feed.
30	11.24	5.6	1/16	¼	Rough	......	....	2	⅛ in. tetal feed. Belt slipped. 8 r.p.m.

utes' run; chips size of small corn grains; tool steel burned, but drive O. K.; belt slipped; tool post shaky; sparks from tool; have to reduce speed to prevent chatter; feed broke on this cut; casting so hot could just bear hand on it; chips 1/8 x 1/8 x 3/32 inch stalled the motor, etc.

From these notes the superintendent can make out a capacity table for each machine. It will be a table giving the maximum that the machine can safely stand. It will be the output measuring stick. For instance, the capacity table for the above cylinder boring lathe would look like Table III.

The explanation of Table III is as follows:

Table III—Capacity Table for No. 98 Drapier Boring Lathe
12 Cubic Inches per Minute—Speeds, 28-48 Feet per Minute

Diam in.	Rev.	Ft.	Cut 5/8 in. Deep		Cut 1/2 in Deep		Cut 3/8 in. Deep		Cut 1/4 n. Deep		Cut 1/8 in. Deep	
			Feed	Cu. in.	Feed	Cu. in.	Feed	Cu. in.	Feed	Cu. in.	Feed	Cu. in.
30	4½	35	1/32	8	1/16	13	1/16	9½	⅛	13½	¼	12½
24	4½	28	1/16	13	1/16	10½	1/16	8	⅛	10½	¼	10½
22	8½	48	1/32	11	1/32	9	1/32	7	1/16	9	⅛	9
18	8½	40	1/32	9	1/32	7½	1/16	11	1/16	7½	⅛	7½
14	8½	33	1/32	7½	1/16	12	1/16	9	⅛	12	¼	12
10	14½	38	1/32	9	1/32	7	1/16	11	1/16	7	⅛	7
8	14½	30	1/32	7	1/16	11	1/16	8½	⅛	11¼	¼	11¼
7	26	48	1/32	11	1/32	9	1/32	7	1/16	9	⅛	9
6	26	42	1/32	10	1/32	8	1/16	11¾	1/16	8	⅛	8
5	26	34	1/32	8	1/16	12¾	1/16	9½	⅛	12¾	¼	12¾

Spindle speeds, 234-14½; 132-8½; 80-4½; 47-2½; 26-1½.

The table was made for the No. 98 Drapier boring lathe covering all the feed and speed combinations that could produce a cutting output of approximately 12 cubic inches per minute. Similar sheets were made out for 10 cubic inches, 8 cubic inches, etc. These were referred to in making cutting tests to avoid figuring each time. Taylor uses a slide rule for this. These tables are used instead of the slide rule. On each step of each cone on each machine in the plant is stamped the spindle speeds when the belt is on that step. This is to get rid of using a watch and is used by the workmen in selecting speeds. A single back geared machine has two of these numbers on each step of the cone.

The Drapier boring lathe cone steps are stamped 234-14½, 132½-8½, 88-4½, etc. These are the revolutions per minute that the spindle of the machine runs. When the belt is on the 132½-8½ step the spindle runs 132½ revolutions per minute with the back gears out and 8½ revolutions per minute with gears in. These speed figures are put at the bottom of the table for the Drapier boring lathe, as shown above.

Holding Machines to Their Work

This seems like a lot of work, but there is nothing the superintendent can do that will prove so profitable to the firm. It will bring him face to face with the fact that the output of his plant is mighty poor. The foremen will get interested in the output at the point of the tool and become faddists on the subject. The workmen soon learn what the correct cutting speeds are.

A plant is like a family with a lot of children. All cannot become great producers. Some are weaklings. The machines are the children of a plant. They all have different capacities for work, and no fixed iron-clad rule can be laid down for all alike. A machine having the power of drive and power of feed capable of taking off 12 cubic inches of cast-iron chips per minute should be held up to this rate. Not only the cutting capacity of each of the older machines, but the cutting capacity for each step of the cone on each machine has to be found by test on the machines regular work, and a table made out for the machine. Cutting capacity (cubic inches of iron removed per minute) is more important than cutting speeds. It is possible, even on an old, weak machine, to take a cut at over 1000 feet per minute if the amount of metal removed per minute is kept small. Nothing counts but cubic inches of metal removed per minute.

Superintendent vs. Machine Salesman

A great thing about the superintendent's taking up this work personally is that it puts him in a position where he knows absolutely what to specify when he orders new machine tools. New tools are not as yet up to the capacity of the tool steel, except in a very few cases. Any one who buys a heavy cutting tool that is not able to burn the tool steel easily is buying what will be an obsolete machine in a few years. One cannot depend on what the tool salesman says, yet, if the superintendent is not an authority on cutting speeds, feeds, cubic inches per minute and capacity of modern tool steels, how can he intelligently select a machine tool to do his work?

The machine salesman will tell him that a feed of 5 inches per minute is the practical limit on milling work. He will try to sell a machine to run at a cutting speed of 60 feet per minute when it ought to run 75, 80 or 90 feet on cast-iron milling. His reason is that he knows that his machine lacks the horsepower to take the heavy feeds and speeds.

The salesman will say that this is what certain other shops, which he mentions, are doing. Now here is just the point: The firm should be turning out 20 or 50 per cent. more on their machine tools than these shops.

ARTICLE XVIII

CORRECT SELECTION OF MACHINES

The Economical Range of Production for Which Each Machine in the Shop Should Be Adapted —What Limits Machine Production

THE different operations in a machine shop are few in number. They are the machining off of a flat surface, the turning off the outside of a cylindrical surface, the machining the inside of a cylindrical surface (boring), drilling, tapping and threading. For each of these operations there are a number of styles of machine tools from which to select.

For machining a flat surface the milling machine, the planing machine or the shaping machine, the lathe or boring and turning machine, or the different forms of grinding machines may be used. In general, the disk grinding machine will turn out two to three times the work that the milling machine will do, and the milling machine twice as much as the other machines in a given length of time.

The Field of the Grinding Machine

The disk grinding machine will surface pieces 6 x 6 inches and under, of unrelieved surface, or 36 square inches in actual surface, at a speed of 18 square inches per minute. On larger work than this the grinding pressure becomes so reduced that the work will be turned out more slowly than on the other types of machines that do surface machining.

The grinding machine's field of work is on pieces that may be trued up by taking off but little metal. One of its advantages, over the other forms of machine tools that do surface machining, is that it requires practically no chucking fixtures. The time and expense of chucking the work is eliminated, the work being held on the grinding table by hand. This makes machining practically a continuous process.

For small faces that are to be hand scraped, inexpensive little emery wheels are available. These are equipped with a longitudinal and cross slide carriage that will finish the work as true or truer than can be done by hand scraping, and in one-third the time. Such a machine will pay for itself in two or three months. For larger surfaces expensive cup-shaped wheel grinding machines of great output capacity and low operating cost per piece, are on the market.

Advantages of the Milling Machine

The milling machine is a rapid producer because cutting speeds of 70 to 90 feet per minute can be used where 60 or 70 feet per minute is the limit of a lathe or boring mill. Several cutters are working at the same time, compared to one, or at the most two, in other forms of facing machines. One piece can be chucked and unchucked while the machine is working on another, making the machining practically continuous. Two or three faces of a piece can be machined at one pass on a planer type milling machine with a number of heads. With this style of machine the machining cost is extremely low.

High cutting speed is possible on the milling machine because each cutter cuts only part of the time, and the cutter edge has a chance to cool between cuts. The milling machine has an enormous output when used with the cutter head or facing head. It will operate with a linear feed of 10 inches per minute on cast-iron surfaces on coarse work and 4 inches per minute on fine work, such as surfaces that are to be hand scraped. The lathe or boring machine can not compete with the milling machine feeding at these speeds.

The spindle of the milling machine should be at least one-third the diameter of the cutter head. A weak spindle will spring, making an untrue milled face, unless the rate of feed is reduced. This is especially noticeable when machining a face that is partly interrupted and partly solid and broad. If the cutter head were backed up near its edge with a bearing shoe the maximum speed and feed could be taken with no danger of the spindle springing. The spring back of the cutter head would be no greater than the down spring of a planer table.

A feed of 10 inches per minute can be maintained with a stiff spindle A feed of 2½ inches per minute only can be maintained if the head and spindle springs. This is a 400 per cent. difference in the output of the machine.

Different Styles of the Milling Machine

The vertical milling machine is best adapted to small and flat work. If the surface milled stands high above the table, the horizontal pressure of the cut throws a heavy strain on the chucking fixtures or strapping-down bolts. High work tends to lift the table from the bed, and this produces chatter.

The horizontal spindle machine is free from the above faults. It is the ideal machine for fast and heavy cutting, as all the strain comes straight down upon the table. This strain, when cutting steel, may run as high as 100 tons on a powerful machine doing heavy work.

The milling machine table should be 6 or 8 feet long. The long table keeps the wrench, used in strapping down the work, away from the revolving cutter head, and gives the workman plenty of room to chuck and unchuck pieces while the machine is on other work, thus getting a continuous output from the machine.

In handling a machine under the continuous system of cutting it is better to do the work in two sections, one at each end of the table. The machine should mill the piece or pieces at one end while the workman is chucking and unchucking at the other. Having a piece or pieces at each end of the table is better than placing them in a continuous line. In this way the operator will not drop into the habit of waiting for all the pieces to be finished before chucking a new lot. He will not hold the machine idle while he chucks these.

The advantage of the above method of arranging the work, instead of filling the table in a continuous line from end to end, is shown in the following example of work done on a Brown & Sharpe No. 5 milling machine. Pieces with a 5½ x 6 inch face were to be machined by a cutter head 7½ inches in diameter, running at a speed of 40 r.p.m. This gave a cutting speed of 77 feet per minute. The depth of cut varied from 0 to ¼ inch, or an average of ⅛ inch. Each piece required 2 minutes when arranged in a continuous line, as against 1¼ minutes when chucked in two groups, one at each end of the table. The two operations may be analyzed as shown below. Two men, a machinist and a helper, operated each machine.

Continuous line milling:

0 min.	0 sec.	Start cut on 7 pieces.
4 min.	25 sec.	Helper starts unstrapping.
6 min.	25 sec.	Both wait for cut to finish.
7 min.	15 sec.	Cut finished. Start taking off the three remaining pieces and brushing off the chips from the table.
9 min.	15 sec.	Pieces all off, still brushing off chips.
9 min.	55 sec.	Start placing seven new pieces.
10 min.	40 sec.	Square pieces up with a straight edge.
11 min.	55 sec.	Both men pull straps down tight.

13 min. 10 sec. Pieces secure. Run table back and feed up to take cut, using index dial on feed screw to set machine for cut.
13 min. 25 sec. Start cut.
Cycle complete. 1 minute, 55 seconds on each.

Below is the cycle of operations analyzed when machining the same job one piece at each end of the table:

0 min. 0 sec. Run table to right end by power rapidly. This mechanism was put on the machine after it was installed in the shape of two sprocket wheels and a chain belt.
0 min. 10 sec. Feed in casting to correct depth of cut. Set this by micrometer dial.
0 min. 25 sec. Cut started on left end casting. Operator and helper change casting at right end of table.
1 min. 25 sec. Cut finished.
1 min. 30 sec. Run table rapidly to other end by power.
1 min. 45 sec. Cut started on casting at right end.
2 min. 35 sec. Run table to right end.
Cycle completed. 1 minute, 17½ seconds on each casting.

Cutter Heads

The cheapest way to make a small cutter head or facing head is to machine slots in the edge or circumference of the head, drive in steel cutters and peen the metal in the cutter head along the side of the cutters. All cutters should be put in at an angle so that the chip will be lifted from the work—not scraped off as is the case where the cutter is put in square. The easiest way to cut the slots is to first drill holes where the ends of the slots are to come, and then plane from the edge back into these holes. Afterward fill the holes by driving pins into them.

Avoid the all geared headstock lathe. It has too much machinery in it. The three-step cone pulley head lathe with a two-speed countershaft is better even for electric drive because it costs less and is more durable.

The modern powerful engine lathe equipped with cross-feed and longitudinal stock will turn work out so fast that the operator will be kept busy taking pieces out and putting them in.

I have seen the work come from such a lathe so hot on account of the high cutting speed that the operator had to wear cotton gloves.

Working with two arbors and two lathe dogs he would barely have time to press out one arbor from a piece, put it into another piece and put on the lathe dog during the time the lathe was doing the machine work on a piece on the other arbor.

The Gang Drilling Machine

Gang drilling machines are gigantic producers. So much so that they are used for all sorts of machine operations that in former years would not have been thought suitable for a drilling machine. 650 bushings 3 inches long, having a 15/16-inch hole through them can be bored, reamed and faced in a day. 400 bushings $4\frac{1}{2}$ inches long, having a 2 7/16-inch hole through them can be bored, reamed and faced in a day.

All the 32 machine operations on 70 pump cylinders, such as are used on the tank wagons of threshing engines, can be done in a day by a man and his helper on a gang drill. This shows what a gigantic output such a machine has.

Gang drilling machines are suitable only where duplicate pieces are turned out by the thousand because the expense of jigs, fixtures and tools runs very high. A four-spindle gang drilling machine generally requires four or five expensive jigs for each individual size of piece turned out. This means four or five times the jigs and tools that a regular drilling machine would take on the same work.

In most plants the gang drilling machine stands idle because duplicate work enough cannot be found to pay for the expensive equipment of jigs, fixtures and tools.

Limit Machine Production by Strength of Piece Worked

A machine should be powerful enough to do the work rapidly. The point that limits the output of a machine tool should be the strength of the piece worked on after it has been strengthened to resist a heavy cut. It should not be the power of the machine tool or the strength of the chucking apparatus. The work should be held in a fixture so secured and the machine should be so powerful, the feed and speed of the cut should be so great that any increase would distort the work or break it.

This rule secures the cutting limit, but how little is it being followed! The writer recalls a chucking lathe that was boring a 2-inch hole in a steel bar, using a Celfor drill. The feed was 3/8 inch per minute. The correct feed for this size of drill in steel is 3 inches per minute. On speeding up the machine and increasing the feed, the belt slipped. By tightening the belt the feed was brought up to 5/8 inch per minute, but the strain was so great on the machine that the teeth on the back gear tore out. The 5/8-inch feed was too heavy for the machine. The 3/8-inch feed was about the limit of its capacity.

Here was a machine whose output was only one-eighth of what it should be. The output on this operation could have been increased eight times. This is not an increase of 25 per cent. nor 50 per cent. but 800 per cent. If one of the office employees worked one hour per day and demanded eight hours' pay he would be thrown out. Yet that is what the chucking lathe was doing every day. Make the piece worked on decide the limit of the speed and not the machine tool. In this case it was the machine tool that settled the speed of output.

The limit of speed at which a piece of work can be machined is probably from ten to twenty times faster than is done on the average. The limit is unknown. Take, for instance, the boring of a 22-inch cylinder. One roughing cutter in a boring head, cutting 35 feet per minute with a depth of 1/8 inch on a side and 9/32-inch feed, will remove 14.72 cubic inches of metal per minute.

Put in six cutters, each with a feed of 9/32 inch, or a total feed of 1 11/16 inches per revolution. If the cylinder is 15 inches long, the roughing cut would be taken in one minute. This is ten to twenty times faster than is done on the average, and shows what is possible if we go to the limit. The casting might have to be straightened and well supported in a jig. Assuming 1/2 hp. per cubic inch of metal removed per minute, this machine would require a 45-hp. motor. The drive and feed gearing, the whole machine, in fact would have to be designed to take care of this power.

The same proposition is true on the milling machine. Say 10 cubic inches per minute is a conservative estimate of the capacity of a single cutter. With cutters located close together in the cutter head so that six cutters would cut at the same time, 60 cubic inches per minute would be the output. Sixty cubic inches per minute means machining a face 10 x 48 inches to a depth of 1/8 inch each minute. This would take a 30-hp. motor and a machine built in proportion. If the milling machine were built strong enough and the piece worked on stiff enough to stand the strain of the cut, the above would be practical; that is, the tool steel would stand this cut. Are many doing this? No. That is why I say the average output can be increased ten to twenty times.

We do not know what the limit of drilling speed is. In a test, on cast iron, a 1 1/4-inch drill drilled 30 inches deep in one minute. This was the limit of the drilling machine, but not the limit of the twist drill.

ARTICLE XIX

MACHINING OF CYLINDERS

Proper Sequence of the Various Operations—Advantages of Different Types of Drilling Machines—A Cheap Tapping Machine

WHEN machining cylinders, it is best to mill them first and bore them afterwards. The following is a good sequence of operations: Catch the cylinders on the flanges in a line of chucks on the table of the milling machine. The chucks could be similar to the one shown in Fig. 24 so that the cylinders would rest on four fixed points, two at each end of the cylinder. These bearing places catch the under side of the flange about 45 degrees on either side of the vertical axis. Mill off one end of the cylinder. Chuck the flange end against an angle block and mill the other end. Finally mill the remaining faces.

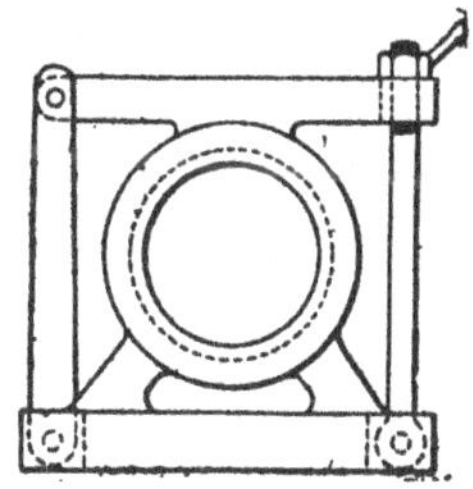

Fig. 24—Jig for Holding a Cylinder During the Milling Operation

If there is enough duplicate work to make it pay, the cylinders should be milled in a machine with a cutter head on each side of the table and possibly with the third cutter on top. One pass will then finish both ends and one side of the cylinder. Such a machine would cost $10,000, however, and the firm should have enough work to keep all three heads going all the time in order to make it pay. No matter what kind of machine is used, the operator should take off the machined cylinders and put on the rough ones while the machine is cutting, thus making milling a continuous operation.

Boring the Cylinders

From the milling machine the cylinders go to the boring machine. Several methods may be used for this process. The cylinders can be chucked in a lathe and revolved, the boring being done with a stationary bar and cutter. They can be bored four at a time in a fixture on the carriage of a standard high power engine lathe. The lathe should be rigged with four parallel boring bars geared to the lathe spindle. They can be bored in a special boring lathe, having two spindles that bore from one end The carriage of this machine has a turntable with bolted-on angle plates against which the ends of the cylinders are strapped, the castings having been milled before boring. While the machine is boring cylinders at the headstock end of the turntable the operator is removing the bored cylinders and putting on rough ones at another point on the turntable. This gives a maximum output from the machine and operator with a minimum pay-roll expense.

One man on the milling machine and one man on the boring lathe can turn out 100 4 or 5 inch cylinders per day ready for drilling. A small floor area thus is very productive. This special boring lathe in some cases is made very elaborate by having a milling head travel across the end of the cylinders at the tailstock end of the table. The machine then bores and mills the cylinders at one chucking.

Using the Drilling Machine

A powerful three or four-spindle heavy gang drilling machine might be used, the cylinders being set on end on the machine table in a suitable set of chucks for boring. With this outfit the output would be controlled by the speed at which a man can chuck and unchuck cylinders and change cutters. If the cylinders are cast in pairs, they will be bored one casting at a time. If the cylinders are single bore castings, two castings are bored simultaneously.

Cylinders can also be bored vertically under very heavy drilling machines made for this work. Three single machines, side by side, run by one operator, will turn out more work and cost less than the two double special vertical machines usually bought.

The jig may be arranged so that after boring one side it can be moved over a fixed distance to bore the other cylinder. This style of boring machine allows the workman to chuck and unchuck the work while the machine is cutting on other casting. The number of machines that one man runs can be increased until the maximum output is obtained from the man. Vertical boring has the advantage over horizontal boring of letting the hot chips drop clear. This keeps the cutter cool.

Advantages of Different Machines

Machine tool makers have not taken full advantage of the improvement in tool steel, especially as to drilling machines. Few radial drilling machines will drill to the destructive limit of a $2\frac{1}{2}$-inch drill in cast iron, say, a feed of from 15 inches to 30 inches per minute. Such a machine would do all the work for a large plant, one man running it and a second man bolting down the pieces and removing them. The drilling cost on a piece would be practically nothing. Many pieces under such circumstances would be drilled rather than cored, as drilling would be cheaper than coring.

A radial drilling machine will turn out work nearly twice as fast as a machine with a fixed spindle, because the moving of the drill from one hole to the next is done quickly on the radial tool and very slowly on the other style.

For all small drilling, say 3/8-inch holes and under, use a fixed spindle, light, sensitive, hand-feed drilling machine. On a sensitive machine the operator can feel when the drill catches and can prevent its breaking. From 3/8-inch holes up to and including about 3/4-inch holes there are powerful little radial drilling machines on the market, which can be quickly handled. They have a large box table so that the operator can be strapping and unstrapping his work while the drilling is going on. This machine will drill a 1/2-inch hole 20 inches deep, a 3/4-inch hole 12 inches deep, or a 1-inch hole 6 inches deep in 1 minute in cast iron. For 3/4-inch holes and larger, and for pipe tap work, use a powerful radial or a powerful fixed spindle drilling machine with a table that can be moved across under the spindle in two directions.

The combination of a radial drilling machine set next to a horizontal one having a car with a turntable on it will save about three-quarters of the chucking and handling time on all large work if the car track runs from the horizontal drilling machine to and under the radial tool.

The horizontal machine will drill all the holes in the sides of a casting and the radial will drill holes in the top. The casting is strapped down to the turntable and revolved to present the four sides to the horizontal drilling machine.

Tapping the Cylinders

A cheap tapping machine can be made on the end of a double hinged arm attached to a post. The arm is hinged like a jointed wall gas fixture, so that the end can be moved over any point of a surface to be tapped. The post end will be attached to a heavy vertical bar of finished shafting so the machine can be set at any height to correspond with the work. The tapping machine, at the swinging end of the arms, is driven from the ceiling through a telescoping shaft, knuckle jointed at each end. This shaft is made from a square bar of key steel, which slides in an iron pipe with lead poured around the square shaft. Such a machine will do the work of three or four hand tappers.

It is sometimes found better to tap the holes on the radial drilling machine at the time of drilling, for this saves an extra handling of the pieces. Studding may be done on this machine at the same time that tapping is done, although not always, as the exact length of the studs may not be known. Quick change collets, or couplings for drills, taps and stud drivers, speed this class of work.

ARTICLE XX

BUSINESS MAXIMS

How Buying for the Factory Should Be Conducted—Attitude Toward Selling and Management of Salesmen—Fundamentals in Advertising

IN every plant money may be saved by obtaining bids on all material bought; a difference in prices is always found for the same grade of material made by different firms.

Never buy in excess of your needs. Over-buying is a common fault and a bad one. Every dollar's worth of unnecessary stock represents idle capital unnecessarily risked. It would be better to place the money that is uselessly tied up in rough or finished material out at interest. Let it bring in an income. A hand-to-mouth plan of ordering is best, provided that sufficient material is always on hand or on order to keep the machines supplied. An idle machine earns no money for the plant.

Order so that you will receive a small, steady stream of material for the machines, just enough so that the machines will keep the assemblers busy making finished product. Have no stock on hand in the rough or finished state except that which is necessary for work moving through the plant. This insures all money spent for material and pay roll showing up in finished machines.

Watching the Cost of Material in Stock Bins

A 7/8-inch nut costs one cent. A bin full of 7/8-inch nuts is the same as a bin full of pennies. Look at your piles of stock on hand in this light. Go from bin to bin, shelf to shelf, and rack to rack, with a pencil and paper, and estimate the cash value tied up in each. You will then begin to see where you can cut down this tied-up capital. Let the firm from whom you buy carry your stock.

A system used by some firms is to mark the material bins with the maximum and minimum amounts of each class of material, rough and finished, to carry in stock and the amount to order. This is a good practice, and it prevents too much money being tied up in unused stock. The clerical labor necessary to carry out this scheme, though, should be watched, so that the pay-roll is not increased. If you have to put on an extra man to look after these maximum and minimum quantities, he will cost from $700 to $1000 per year. A lot of stock could be bought for less than this amount. Every added expense of this character should be made to pay for itself, and to yield a profit in addition—not a bookkeeping profit, but a profit of real dollars at the end of the year. If you increase the pay-roll to take care of the stock, you must, by means of this added expense, be able to increase the profit of some or all of the manufacturing departments. If this cannot be done, the change is not worth the expense.

Manufacturing Finished Machines Not Finished Parts

Run your product through in exact lots. If you decide to make 25 machines of a certain size, get out the material for all the parts of the 25 machines down to the last small piece. Run no more parts through than are enough for the 25, no matter what the temptation may be, to run more of any one piece. If some parts have to be scrapped, start through enough more to bring the number up to 25.

Besides the advantage of the great reduction in the running capital needed to carry on the business, there will be an enormous gain in space that was previously occupied in storing finished parts. This will give room for putting on more men, and thus increase your output, and likewise your profit, with no additional buildings or ground space.

Keep in mind that the object of manufacturing is to make finished machines, not finished parts. Not long ago the author received from an establishment that had failed a list of material that it had on hand of which it wished to dispose. It showed that the firm had violated this rule. They had an overstock of everything.

The Selling Department

A firm must have a persistent determined selling system which will dispose of its full plant capacity at the least expense. The head of the selling department should not spend too large a proportion of his time at correspondence with customers, or actual selling, but should occupy himself in establishing more and better agents and dealers, and in writing to them often. He should see that the amount of goods which they sell is in correct proportion to the population of the district. Each agent and dealer should receive a letter of some sort, about selling, from the sales head every few days.

The firm should have a system of keeping track of the daily operations of its traveling men. It is well to send out three traveling men, first giving them complete instructions on all the selling points of the product, with the route of each laid out. An expense and sales record of each man is necessary in order to ascertain the percentage relation between his expenses and the value of his orders.

After three or four months' trial, the traveling man whose orders cost the most to get may be replaced by a new man. A schedule can be made in time giving the rate of improvement a new man must make to hold his job.

The standing of a salesman should depend upon the value of his sales as compared with the business population of his district. The salesman who has the best district should send in the greatest value of orders. A well-established district will make selling easy. A man who is placed in a district from which few orders in the past have come should not be expected to make the sales of one who is in a well-established district.

A salesman should be kept in one district as much as possible. He becomes more valuable as he gets acquainted with the people in the district. People will listen to a pleasant fellow the third time he comes around, when possibly they won't the first.

Questions for the Traveling Man

The traveling man should ask the following questions of the possible customer, if he can work them into his conversation without causing offense:

1. "What trade papers do you pay the most attention to?" The object of this is to find what papers are best to advertise in.

2. "What time of the year do you generally do your overhauling or buying?" The object is to find out when to go after this particular man's business; when is the best time to send business-getting letters and traveling men to his style of business. Selling to be done right, should be handled systematically. The salesman, the advertising literature, and the trade paper advertising, should be sent forth at the time when each dollar spent will bring in the greatest return.

3. "What do you pay for competitors' goods?"

4. "What objections have you to our product?" These objections will be entered in a book kept for this purpose. A convincing answer will be thought up for

each objection and entered in the book. Make a correction in the design of the product, if there is any foundation for criticism.

5. "What objections have you to our competitors' make of machinery;" Enter these in the book. Salesmen must learn the contents of this book, for it will contain a good series of selling arguments.

Routing the Traveling Man

Use the map and tack system for routing traveling men. Divide the country into districts, with a very large city as a center for each district. Let the traveling man for the district live in the large city. This will give a low cost of selling in the large city, with no hotel expenses, and with short trips. Credit the salesman for all orders that come from his district. Each district must be charged with a pro rate to cover advertising, catalogues, circular letter writing, and expense of head sales manager. The pro rate charge will be proportioned to each district in accordance with the density of the buying population of the district.

The district where the factory is located will be the educational district for new salesmen. Better weeding out can be done here, as the new man is under the immediate eye of the sales manager.

There are wonderful salesmen in the world; geniuses in their line; men that can sell anything. A firm must never be satisfied with those who barely make good, but must keep trying until their men are all wonders.

The sales manager should instill enthusiasm into traveling men by frequent talks. When on the road they should receive a letter daily from the home office. A letter received by a man each day about his work is the same as a foreman coming around looking over the work. Nothing will boost him along like these letters;

they prevent discouragement. Dealers should also be written to often for the same reason, as stated.

The names of the firms that are enormous buyers should be known, and the amount of business they probably could give, and also how much should be spent each year in order to obtain their business. Some of these concerns may buy enough to make it pay to have a man constantly sitting on their door step.

Selling Just as Tangible as Manufacturing

Selling must not be considered in a hazy sort of light. It must not be considered a case of luck that orders happen to come in. Selling is just as tangible a business as manufacturing. In proportion to the number of salesmen out, agents taking the goods, advertising, etc., will be the volume of business taken in. If a manufacturer had to double his plant output, he would have to double the force of workmen in his plant. The same is true of selling. Four salesmen will sell twice as much as two, provided they are good salesmen and are well directed, and salesmen have to be directed the same as producers in the shop. The sales manager has to give his full time to directing. The sales manager is a foreman over the salesmen. Work would soon be turned out at a loss in the shop if the men were not directed. The same is true of selling.

Like manufacturing the selling must be directed on the most economical lines. The money spent for advertising must be spent where it will bring in the greatest return. Each year a firm should have more and better dealers than they had the year before, and more and better salesmen. If a firm does not reach out in the selling department, it will never be able to reach out in the manufacturing department.

Circular letter writing should be pushed to the limit. The letters should be written on the regular paper that the firm usually uses, not on a cheap grade of paper. A large mail sack of letters should leave the office every day.

The selling price of the product must be as low as that of competitors. To make a profit, selling at this price, the product must be of an inexpensive design to build, the most inexpensive manufacturing methods must be used, a maximum output per man must be obtained, the overhead expense must be kept down, the material must be bought at lowest possible prices and the stock on hand must be kept at the minimum.

The Personality of the Salesman

The salesman must have energy and enthusiasm. He must be of a very pleasing personality. He must be a kind of person delightful to have around; a kind of man whose arrival the buyers look forward to. When the best salesman comes into an office, generally all work stops. He is such a delightful person, such a good all-around talker, that all hands sit down and have a regular talk feast. Half the battle is won if the customer is pleased to see the salesman.

I remember one of the largest and oldest lathe builders in Cincinnati related a little incident about selling, which illustrates this point of a pleasing personality in salesmanship. A man, who in former times had been on bad terms with the Cincinnati lathe builder, was in the market for twelve lathes and swore that he would not buy a lathe from his old enemy under any consideration. The Cincinnati lathe builder sent his best salesman to see this antagonistic customer. The salesman brought back the order for the dozen lathes. The lathe builder asked the salesman how he managed to get the order. "Tell me just what you said and what he said. Give me the whole conversation. Didn't the buyer show great antagonism to us." The salesman said: "Yes. He came out to see me and started in a regular tirade against our firm and our lathes. Said he would let his plant rot down before he would buy one of our lathes." The lathe builder asked what the salesman did then. The salesman said: "I just laughed." "What did he do then?" asked the lathe builder. "He just laughed," said the salesman. It was the laugh that sold the lathes.

The Points to Be Made in Advertising

The object of advertising is to instill into the mind of every possible buyer these four things: The firm's name; the location of the firm; the class of goods manufactured and sold; and why they are more desirable than other firms' goods.

When getting up an advertisement the advertiser should decide what proportion of the space to give to each of these four points. That point should be made the most prominent which is the hardest to impress on the reader; that point in which he is interested the least. This is undoubtedly the firm's name. No matter how good an advertisement be, it is a failure if the firm's name is not impressed on the reader. A reader will look through pages of advertising and never have a firm's name impressed on him. This is the one point in which he is least interested.

A manufacturer becomes so familiar with his own name that he is liable to overlook the fact that he is unknown to the great majority of people. If he doubts this, let him ask his traveling men whether or not they find people who have never heard of the firm. The traving men are coming in contact with such people every day.

The author remembers his general foreman saying of an advertisement: "That's a fine, catchy advertisement." When asked whose advertisement it was, he said,

"Oh, I didn't notice. I was only speaking of the advertisement."

Once the author's firm sent out a lot of beautiful little glass clocks, with its name across the faces in small, neat letters. One of the firm went to see a large institution in the state which was in the market for its product. The head of the institution said: "I am very sorry, but we have just ordered. To tell you the truth, I did not know there was a pump manufacturer in our state." The member of the firm then said: "That little clock ticking there on your desk is an advertisement that we sent you." The manager picked it up and said: "Well, do you know, I have had that on my desk for six months and this is the first time I ever looked at the name."

Few people know the name on their office calendar. It is the firm's name that is the hardest thing to impress on the reader of an advertisement. This must then be given the greatest proportion of the space.

The second point is locality, which should take up the least space, because the name of the town is probably familiar to the reader, and is easily impressed on his mind. The advertising space must be economically used.

The third point is what the firm manufactures. A small cut of the product, occupying about one-tenth as much space as the firm's name, will catch the reader's eye and impress this point on him Therefore, cuts should be small.

The fourth point is the descriptive matter. This should be changed often.

Differences to Be Observed in Pamphlets and Catalogues

The pamphlet that is sent out by mail, with the circular letter, should be full of pictures. Reading is mental work. It requires effort on the part of a person to read something in which he is not interested. On the other hand, looking at pictures is a recreation. If the tale can be told in pictures, the busy man who gets the pamphlet will give it a glance at least before it goes to the waste basket.

The catalogue that is sent in answer to an inquiry should be different. It should have both pictures and reading matter. In this catalogue should be given all the advantages and good points of the product.

In all styles of advertising literature the firm's name should be worked into the reading matter as much as possible. It should appear under every cut. For instance, if the engine is built by Jones & Co., under the cut of the 20-hp. engine should be the words "20-hp. Jones Engine." In the reading matter will appear, "The Jones valve gear is such and such. The Jones governor, etc." The idea when getting up advertising literature is to keep in mind that the whole advertisement will be lost if the firm's name is not remembered.

One of the most important points in trade paper advertising is the advertisers' index. It is here that the buyer turns to find the names of the firms to write to for prices.

INDEX

A

PAGE

B

C

D

E

F

P

R

S

T

Spotted something that isn't quite right?

Let us know and we'll fix it for future reprints. And as a thank you for your diligence, we'll send you a free Sothis book of your choice.

publisher@sothis.press

This page intentionally left blank

Made in the USA
Monee, IL
08 November 2026

68119176R00132